The Fire Department Water Supply Handbook

WILLIAM F. ECKMAN

Fire Engineering Books & Videos

> **Disclaimer**
> The recommendations, advise, descriptions, and the methods in this book are presented solely for educational purposes. The author and publisher assume no liability whatsoever for any loss or damage that results from the use of any of the material in this book. Use of the material in this book is solely at the risk of the user.

Copyright© 1994 by
PennWell Corporation
1421 South Sheridan Road
Tulsa, Oklahoma 74112-6600 USA

800.752.9764
+1.918.831.9421
sales@pennwell.com
www.FireEngineeringBooks.com
www.pennwellbooks.com
www.pennwell.com

Director: Mary McGee
Production / Operations Manager: Traci Huntsman

Library of Congress Cataloging-in-Publication Data Available on Request

Eckman, William F.
The fire department water supply handbook / William F. Eckman.
p. cm.
Includes bibliographical references and index.
ISBN 0-912212-35-7
ISBN13 978-0-912212-35-7
1. Fire extinction—Water-supply. I. Title.
TH9311.E28 1994
628.9'252—dc20   94-2217

All rights reserved. No part of this book may be reproduced, stored in a retrieval system, or transcribed in any form or by any means, electronic or mechanical, including photocopying and recording, without the prior written permission of the publisher.

Printed in the United States of America

2 3 4 5 09 08 07 06

The Fire Department Water Supply Handbook

Contents

	INTRODUCTION	vii
1.	An Introduction to Fire Department Water Supply	1
2.	Water Supply Master Planning	28
3.	Risk Analysis In Fire Protection Planning	53
4.	The Use of Fire Hose in a Water Delivery System	67
5.	Testing Water Supply Apparatus	83
6.	Fire Hydrants As A Water Source	125
7.	Suction Supply As A Water Source	165
8.	Relay Operations	212
9.	Water Shuttle Operations	237
10.	Water Shuttle Dump Site Operations	262
11.	Water Shuttle Fill Site Operations	293
12.	High-Capacity Water Shuttle	322
13.	Specifications for Mobile Water Supply Apparatus	343
14.	Standard Operating Procedures	378
15.	Water Supply Officer	393
	GLOSSARY	416
	INDEX	429

Introduction

IN MOST OF THE BOOKS that deal with fire attack and tactics, it is assumed that fire hydrants are available to supply water when it is needed. Why is this not a reasonable assumption?

Where public water systems exist, they may not have enough hydrants, properly distributed, to cover the area effectively. In other cases, the maximum flow rate from the systems may not be high enough to handle the buildings at risk. In many communities the water system has not been expanded fast enough to cope with the development that is taking place. Buildings frequently are constructed before fire hydrants are installed. When hydrants are available, the maximum flow from the system may be inadequate to provide the needed fire flow. Even when the system is well designed and constructed, there area areas that are not accessible to hydrants for one reason or another. Limited access highways and bridges are two causes of inaccessibility.

Rural areas pose even greater problems. Fire protection is usually handled by individual volunteer fire departments with little or no relationship between the needs of the area and the apparatus and equipment that are available to meet them. Some departments have purchased tankers to provide a water supply where hydrants are non-existent or too widely spaced, but there is no standardization and little co-ordination of efforts. Fire departments have been going it alone, trying to build tankers that are large enough to handle fire protection in their area.

While fire flow requirements are generally well defined for metropolitan areas, and hydrant systems are carefully designed and evaluated to meet the needs, little or no attempt has been made to extend the same techniques to nonhydrant areas. There are no generally accepted standards for the movement of water by tankers, and very few fire departments make good use of the static sources available to them. Storage capacity, recovery rates, and access to water supply points all need to be evaluated and improved. The topic of how to do water supply planning in areas not served by hydrant systems has been virtually ignored.

This book will attempt to provide the information that fire departments, and specifically water supply officers, need to provide the same caliber of fire protection to suburban and rural areas that metropolitan areas have enjoyed in the past. It will encourage the water supply officer to get involved in the type of prefire planning that is needed to enable the fire department to get the most out of its available personnel and apparatus, and to ensure an adequate water supply for fire protection throughout its area of responsibility.

One of the primary objectives of this book is to call attention to the need for providing an adequate water supply in rural areas. Many suburban fire departments, which are responsible for fire protection on long stretches of limited access highways and expensive residential structures, as well as commercial or industrial buildings located in areas not served by public water systems, do not have the tankers that would be required to maintain the needed fire flow. For all practical purposes, much of the area these departments serve is essentially unprotected. This book will cover some of the methods that can be used to:

1. Determine the needed fire flow within a fire management area;
2. Demonstrate the inability of many suburban and rural fire departments, and the water systems they depend on, to supply the needed fire flow; and
3. Provide guidance in planning to meet the requirements of the territory the fire department covers.

By setting standards for adequate levels of water supply, this book will enable fire departments to establish a standard response to meet them. It will suggest some standard operating procedures that will help fire departments to get the most out of the available water supply. One fire department, after a complete evaluation, discovered that a farm pond, which it had always considered to be its primary water source, was not very reliable. There were other readily available sources that had the potential to be much better than the one it had always depended on, but it was not until an overall water supply plan was developed that the sources were identified.

From the standpoint of Fire Department Water Supply, the most expensive equipment is not always the most effective. By following the standards that are recommended in this book, fire protection officials everywhere will be able to get the most for their money in providing fire protection to the taxpayers and their contributors. These standards will give fire departments, commercial developers, and the insurance industry a goal to shoot for that is attainable, measurable, and cost effective.

This book is the result of experimentation and evaluation of water supply apparatus in conjunction with more than twenty years of water supply seminars conducted up and down the East Coast from Maine to South Carolina. It is also based on experience gained in developing fire protection master plans and water supply plans for both individual fire departments and entire counties. One of the reasons this book was written is the frustration that the author experienced during 25 years as a volunteer fire officer trying to control fires without a dependable water source. It is probably the most comprehensive coverage of Fire Department Water Supply, especially in areas where hydrants are not available or have inadequate flow, that has been written, and introduces some new concepts in high-capacity water shuttle and automatic relay operations. It is based not on classroom theory and engineering calculations, but on the results of hundreds of practical water supply evolutions that have been conducted in different areas with all kinds of apparatus and equipment. The experience gained in conducting fire protection surveys, drawing up plans for improving potential water supply points, and working with local fire departments in implementing these plans, has provided an opportunity to field test most of the concepts and operating practices that are advanced herein.

To both simplify the explanations of some very complex subjects, as well as to introduce some new ideas, the author will use some terms that may not be familiar to all readers. Some terms are common in certain areas of the country, but unknown in others. If there is any question about the precise meaning of a term used in this book, refer to the glossary of terms.

All types of professionals in the fire protection field can benefit from this book. It should be of special interest to line officers in all fire departments, both paid and volunteer. It will be helpful to them in fire protection planning as well as in helping to establish good operating practices and procedures. It will provide a reference for training organizations, and can be used as a textbook for students. Water supply is an area that is neglected in most officer training programs, and good water supply officers are hard to find. This book is intended to be a practical guide to water supply, one that does not require a college education, or advanced technical knowledge on the part of the reader, to be understood. The data contained herein are based on approximations intended for fireground use and fire department planning, and do not represent precise mathematical calculations. The detailed information that is presented on all aspects of fire department water supply will provide a good background for the fire officer who is assigned this responsibility and will lead to good decisions on the fireground as well as resulting in better prefire planning.

Municipal government officials will find this book of value when planning for fire protection in their area of jurisdiction. It provides a basis for making decisions that will get the most fire protection for the dollars that are available for this purpose. Public utility departments that operate central water systems will find some alternate methods of solving fire flow problems in a more cost-effective manner than the traditional expansion and renovation of hydrants and distribution systems.

Industrial representatives will find some innovative concepts they can use when setting up new installations in areas that do not have adequate public water systems. A number of alternatives are suggested to enable developers to provide for unprotected risks when existing fire protection capabilities are less than what is needed. Members of the insurance industry can find some practical guidelines to use to encourage fire departments to meet the standards for Fire Department Water Supply that they have established in their rating guide.

The Insurance Services Office considers water supply to be the most important single factor in an evaluation of fire protection in a community. Everyone involved in the fire protection field should put the same emphasis on it that the ISO does. This book is dedicated to achieving that goal.

Bill Eckman

The Fire Department Water Supply Handbook

1. An Introduction to Fire Department Water Supply

AN ADEQUATE WATER SUPPLY is probably the single most important element in effective control and extinguishment of fire. Except for certain specialized applications, fire suppression involves applying water to a fire in sufficient volume to absorb heat faster than it is being generated. This cools the fuel below its ignition point. To provide water in a timely manner, there must be a fire station located where equipment can be deployed quickly. The responding engine company must have a pumper and the water, equipment, and personnel to make an effective attack. In many cases, a single company cannot provide the needed fire flow.

Each phase of fire suppression, beginning with size-up and continuing through overhaul operations, requires a supply of water adequate to furnish the gallons per minute (GPM) needed. The fire department also must have the ability to transport it to the scene of the incident.

One measure of the importance of water supply in fire protection is the emphasis given it by the Insurance Services Office (ISO) when communities are surveyed to determine their Public Protection Classification (PPC). The ISO Rating Schedule, used to evaluate a community's fire protection capabilities, awards more points for water supply than any other single item. Receiving and dispatching fire alarms represents 10% of the rating, the fire department is responsible for 50%, but the water supply constitutes 40% of the total rating! ISO's rating schedule is based on years of fire loss experience, and recognizes the quality of fire protection within a community by establishing guidelines insurance companies can use in setting premiums. Based on past fire losses, the ISO considers a community's water supply vital to keeping dollar losses to a minimum.

With all of the other demands made on fire officers, water supply too often is dealt with as an afterthought, with little attention devoted to it until the need arises. Frequent comments in news reports that "the fire department was hampered in its efforts to fight the fire by a lack of water" indicate that water supply problems occur regularly. Experienced fire officers undoubtedly can look back on fires they have fought and remember the frustration they felt because the fire was too large to be controlled by the fire streams they were able to put in service.

There is a tendency for fire officials to hold the government of their political subdivision responsible for water supply. Since the days when volunteer bucket brigades were organized, community cisterns constructed, and central water systems installed with hydrants located strategically throughout the community, we have learned to depend on local government to provide water when it is needed. This has created a presumption, generally unspoken, that where fire hydrants are not available, the fire department cannot be expected to put out a fire. This is a Stone Age approach to fire suppression. Technology is readily available to provide an adequate water supply to any location. It has been aptly said that there is never a shortage of water for fire fighting. The real problem is the inability of the fire department to get water from the source to where it is needed.

INITIAL WATER SUPPLY

Water supply should be the priority immediately upon arrival at the scene of a fire, even before hoselines can be laid from the hydrant or whatever source will be used. The first few minutes are critical because this is the opportunity to try to control a fire, ideally, when it is between the ignition and flashover phases, and before it reaches the fully developed stage. At this point it is beyond the capabilities of the fire department to handle until it reaches the decay phase and is beginning to burn itself out, having consumed most of the readily available fuel. The success or failure of the extinguishment effort generally will hinge on the fire department's ability to apply fire streams large enough to bring the fire under control within the first few minutes after arrival.

For the most part, those initial fire streams will come from the tank of the apparatus on the first alarm assignment.

To mount an effective attack, a good rule of thumb is to have a minimum of 2,000 gallons available on arrival of the first alarm units. This quantity can be delivered by a pumper-tanker carrying 2,000 gallons, a pumper with 500 gallons and a tanker with 1,500, or any other combination. An initial supply of 2,000 gallons will support an effective interior attack or provide for exposure protection using multiple light attack lines, one heavy attack line or both. This amount should be sufficient to handle most single-family residences, small commercial or industrial occupancies, or to make an initial attack on a fire in a large building where the fire is still confined to a limited space. Experience has shown that more than 90% of the alarms a typical engine company responds to are handled with less than 500 gallons of water. While this is good from a fire loss and property damage standpoint, it can foster a false sense of security. Unless a fire can be extinguished with the tank supply, an external supply will be required to continue the attack. After large numbers of successful extinguishments using only the water carried in the tanks of the apparatus, it is easy for fire officers to become complacent and overconfident, and subsequently be unprepared for fires where large amounts of water are needed.

NEEDED FIRE FLOW

If the initial attack is unsuccessful, or if the fire is in an advanced stage when the first engine company arrives, additional water will be needed. Rural fire departments tend to think in terms of small lines and limited flow, but the amount of water required to extinguish a given fire in a rural area is identical to that needed in a more densely populated area. Large fires need large fire flows for rapid extinguishment. While exposure problems tend to be less severe and traffic control is less troublesome, there are other complications that make rural firefighting even more difficult than it is in urban areas.

Churches, commercial and industrial buildings, and barns and other large structures located in rural areas present special problems. They are valuable, and those with a high fuel load are difficult to protect. Dairy farms or horse stables contain valuable animals and have expensive machinery. Trucks, construction equipment, farm machinery, and other vehicles may be parked next to or stored in separate buildings close to the fire, thereby presenting an exposure problem. These buildings may be more valuable than their replacement cost

Figure 1-1. Attack Pumper equipped with a 1,000-GPM pump and 1,000-gallon tank.

Figure 1-2. Direct Water Supply from the pump on a Tanker to the Attack Pumper.

Figure 1-3. Target Hazards call for water supply pre-plans.

would indicate. They are essential to the business owner's operation, and, in many cases, losing them could initiate a bankruptcy. This affects the community through loss of jobs, and interruptions in services.

In addition to the fire spread that can be expected once the building becomes involved, the fact that the alarm very likely will be delayed (because it is in a rural area) handicaps the fire department. Once alerted, the fire department's travel time may be slowed, extending the response time. In isolated areas, manpower and equipment tend to be limited. Hydrants generally are not available and static sources may not be accessible to fire department pumpers. Given all of these factors, it seems likely that any fire in a large building anywhere will require a large volume of water for control and extinguishment.

Target hazards frequently are located beyond the effective reach of fire hydrants. High fuel loads, large buildings, hazardous materials, or potential life hazards can be found anywhere, but they pose special problems in suburban and rural areas. Regardless, property owners have a right to expect that an adequate level of fire protection will be provided when needed, including a reliable water supply, which the fire department has a responsibility to deliver. No matter how a fire department is funded, it has the same obligation to the citizens who support it. It has been said that the only time volunteer firefighters volunteer is when they join the department. After becoming a member, a volunteer assumes the same obligations that paid firefighters have—to protect the life and property of anyone who is in need.

AREAS OF RAPID GROWTH

Rapid growth in many areas of the United States, coupled with too little land for a burgeoning population, has resulted in what are called interface areas (areas that formerly were suburban or rural in nature). In some cases interface areas fill the space between cities, melding them into one large area. In this process, new construction often outruns the water system. Public utilities have depended on federal grants or loans for water main construction, but with cutbacks in government funding for financing improvements, many water systems are not keeping pace with new development in the areas they serve. This trend is not likely to change, and the fire service needs to prepare for it.

In many metropolitan areas, new construction has outdistanced the water system. With land values increasing, buildings are constructed in areas that were not considered buildable in days gone by. Water lines that were adequate when they were installed originally may be too small to cope with the demands brought by new construction. New buildings tend to be much larger and present a more severe fire protection problem than the older buildings the water system was designed to serve. Some areas have no coverage because no one anticipated the need. Pockets of unprotected or inadequately served territory can develop in the midst of, or surrounding, a good hydrant system. When this happens, it frequently is more cost effective to provide a supplemental water supply than to go to the expense of replacing old lines or constructing an extensive system of new water mains.

LIFE HAZARDS

The housing boom has led to the construction of high density housing units that overwhelm the water system. Whether the housing stock consists of garden apartments, mobile home parks, or townhouses, it presents special life safety problems. Cost-conscious developers tend to comply with the local codes, and nothing more. When fire safety devices or construction safeguards are required (e.g., smoke detectors or egress requirements), they generally focus on life safety, and will not provide an appreciable amount of help to the fire department in attempting to minimize property damage. Some suburban and rural areas have no building codes, and unsafe conditions may be built in at the time of construction.

Water supply becomes even more problematic when high-hazard occupancies such as hospitals, nursing homes, churches, warehouses storing dangerous materials, or large commercial buildings are built outside of areas served by public water systems. Due to the nature of the buildings, they are required to meet life safety standards as established in local building codes or in state-wide regulations. The standards set by these codes are only the minimum that is acceptable and are not necessarily intended to control fire loss.

Public assembly buildings generally must have sprinkler systems for rapid control of the fire and to allow safe egress for occupants. In rural areas, there will have to be enough water in storage to supply the sprinkler system, but this may not be enough to provide the needed flow for handlines or exposure problems. The fire department needs to be prepared to provide a supplemental supply to allow the sprinklers to continue to operate and for other firefighting needs if the fire progresses beyond the initial stage.

TRANSPORTATION HAZARDS

Modern transportation systems carry all kinds of dangerous cargo. To respond to transportation incidents the fire service has trained hazardous materials (HAZMAT) teams and provided specialized equipment for their use. Sometimes, however, fire departments are unable to provide a water supply adequate for HAZMAT teams to operate effectively, a problem often overlooked in preplanning.

Hazardous materials being transported do not have to be exotic chemicals or materials to create a potential catastrophe. Common materials such as liquefied petroleum (LP) gas, primarily used for cooking and heating, and ammonium nitrate, generally used in fertilizer, have been responsible for some of the most spectacular fires, with an accompanying loss of life and property, in the United States. A fire department may encounter hazardous materials at any call for assistance to a highway accident, train wreck, airplane crash, or commercial or industrial building. When hazardous materials are involved, an adequate water supply, available in sufficient quantity for an initial attack immediately upon arrival and reliable enough to be sustained for hours and even days, will enable the fire department to reduce the life hazard to both civilians and firefighters and prevent the spread of fire.

Figure 1-4. A leaking LP Gas truck requires an adequate water supply to prevent ignition or eliminate possibility of tank rupture if it is burning.

Figure 1-5. Sign fastened to sound barrier on Limited Access Highway to indicate fire hydrant location.

Figure 1-6. Siamese and Gated Wye through sound barrier on limited access highway provides water supply from nearby hydrant.

Accidents frequently occur in places where it is difficult to establish an adequate water supply. Interstate highways are constructed with limited access, isolated from local traffic to prevent accidents. This limited access may isolate the responding fire department from its fire hydrants and makes supplying water difficult, if not impossible. Inner city portions of the highway frequently are elevated above or cut below the streets where the hydrants are located. Entrance and exit ramps complicate traffic patterns and reduce access for emergency vehicles. Even where the highway is at ground level, fences or bulkheads erected for noise reduction make access to hydrants in adjoining neighborhoods difficult. Some cities have installed gates in the fences and doors in the walls and labeled them with the closest fire hydrant location. Some have provided engine supply fittings through walls or bulkheads to allow access to the hydrant system. Even where these steps have been taken, however, providing fire protection on these highways is complicated. One engine company must be dispatched to the incident scene, while another company is required to provide a supply line from the access point to the scene. A third company responds to the closest hydrant and lays a line to the access point. Establishing a water supply in this manner requires coordination and training, and often is too time-consuming to be of any help in critical fire situations that involve hazardous materials or trapped victims.

Bridges also present special problems for fire protection. For example, most bridges are too far above the water to allow a pumper to establish draft. Very few have built-in hydrant systems or horizontal standpipes, so the fire department has to establish its own water supply if emergencies are to be handled promptly and effectively. When a fire boat is available, a standpipe from one of the piers to the deck of the bridge would provide a means of getting the water where it is needed, but when the bridge is several miles long, as many of them are, something more is needed. Life hazards are severe on long bridges, since an accident on the span generally causes traffic to back up, and makes rapid evacuation of the people trapped there impossible. It is important to establish emergency procedures, and develop a plan for each bridge or other special hazard for which the department is responsible. Advance preparation is the best way to prevent life loss and to minimize injuries when an emergency occurs. The fire department must be able to provide an adequate water supply immediately upon arrival.

HYDRANT SYSTEMS

While public water systems, including fire hydrants, are generally the responsibility of the political subdivision or the operator of the system, the fire department has to be involved. Fire officials need to develop a relationship with the utility's management that will enable them to review the system design before it is constructed or before changes are made. While this may not involve actual plan review, it will at least provide an opportunity to make the fire department's requirements known to the engineers who are responsible for any system changes or additions. The fire department should base its recommendations to the utility planners on the needed fire flow for the area to be protected, and the fire code requirements for life safety, in addition to the domestic requirements.

To make sound recommendations for improvements to the water system, the fire department should prepare a water supply master plan for each area it serves. The base fire flow for each Fire Management Area (FMA) should be determined by surveying the types of buildings and risks that predominate. This master plan should include planning target hazards and specific risks to determine how much fire flow will be needed. The available fire flow at the location being analyzed then can be compared to the needed fire flow, and recommendations for improvements made as required. Chapter 3 of this book, Risk Analysis in Fire Protection Planning, covers this topic in some detail.

To ensure that existing hydrants will be available for use, and that they will supply the needed fire flow, they should be inspected regularly and maintained in good operating condition. While the inspection usually will be done by the utility that operates the system, the fire department should not only have access to the inspection records, but should perform a certain number of spot-checks to verify that the work is being done properly. Figure 1-7 provides an example of why the fire department needs to be involved. When the fire department decided to test this hydrant, inspectors found that the barrel was broken and the hydrant was unusable. Although only a 2½"-hydrant, it was the only one within two miles of the location, and the fire department was unaware that it was unusable. To prevent surprises such as this, the fire department should arrange to be notified immediately when a hydrant is taken out of service for any reason. This information should be noted on all plans for affected areas. The Water Supply Officer (WSO) should see that repairs are made and that hydrants are not left out of service for extended periods of time.

Figure 1-7. Broken barrel on hydrant makes it unusable in emergency.

As the primary user the fire department is responsible to see that good operating practices are followed each time a hydrant is used. These practices include opening the hydrant fully, closing it completely after use, and making sure that the barrel drains completely. Any problems with a hydrant should be reported promptly to the responsible authorities so that the hydrant can be repaired before it is needed for an emergency situation.

INITIAL FIRE ATTACK

The National Fire Protection Association (NFPA), in *A Training Standard on Initial Fire Attack* (NFPA 1410), has given the fire service some guidance on initial fire attack. Although this standard covers training, and not actual operations, it establishes some criteria for maintaining an effective water supply, and, while it deals primarily with the ability of the fire department to apply water, it also can be used in water supply planning.

Appendix A to the 1988 edition of NFPA 1410 suggests that the standard response to structure fires should be a minimum of two engine companies. These engine companies should be able to place two initial attack lines in service and back them up with a third line immediately. The two initial attack lines should flow a minimum of 200 GPM with a required flow from the backup line of at least 200 GPM. The time required for this response varies with the type of supply being used.

Figure 1-8. Multiple attack lines require additional water supply.

The standard suggests that when a single large diameter hoseline is used to supply the attack lines from a hydrant, all attack lines should be operating at full flow with the desired pressure within three-and-a half minutes. When the fire department has water tanks that carry more than 500 gallons on the initial response units, it should be possible to get all three attack lines into service in even less time. Table 1.1 converts this expected flow into water supply requirements.

If the fire department arrives on the scene of the incident with 2,000 gallons, as has been recommended, it is obvious from the chart that an attack of the magnitude suggested would exhaust the water in the tank in eight minutes. If the fire department carries more or less than 2,000 gallons on the initial alarm units, this figure would have to be adjusted accordingly. Regardless, the fire department has no time to lose. With a 2,000-gallon initial supply on board, an external water supply would have to be set up and moving water within eight minutes of the arrival of the first engine company if the attack on the fire is to be sustained. If a good hydrant system is available, i.e., with a hydrant located within 1,000 feet of the scene of the emergency, it is possible to meet this requirement. This is not easy—standard operating procedures will

Table 1.1

Time	Fire Flow	Water Used
Arrival + 1 min.	0	0
Arrival + 2 min.	0	0
Arrival + 3 min.	0	0
Arrival + 4 min.	400	400
Arrival + 5 min.	400	800
Arrival + 6 min.	400	1,200
Arrival + 7 min.	400	1,600
Arrival + 8 min.	400	2,000
Arrival + 9 min.	400	2,400
Arrival + 10 min.	400	2,700

have to be developed to ensure that a supply of water reaches the attack pumper with no unnecessary delay—but it can be done. In rural or suburban settings, where hydrants may not be readily available, it is even more difficult to meet this requirement.

It can be fatal to interrupt the water supply to attack lines while an attack crew is inside a fire building and out of sight of the pump operator. During an offensive fire attack there are times when the only protection against death or injury for firefighters on the hoselines is the water from the attack lines. When an attack crew enters an extremely hot room, it is the cooling effect of the fire stream that keeps them from being burned. Interrupting that flow without warning could be deadly. The incident commander should never send hose crews inside a building, or begin an offensive attack, until the water supply has been established, and an adequate supply is assured.

FIRE DEPARTMENT WATER SUPPLY TECHNIQUES

Any time the scene of an emergency is more than 1,000 feet from the closest water supply, specialized techniques must be used. Most engines carry at least 1,000 feet of hose that can be used for supply line. Moving water over longer distances requires the use of more than one engine company (with its apparatus, equipment, and personnel) to establish a dependable water supply. To do this quickly and effectively, a fire department needs both standard operating procedures and an effective fireground organization, and should conduct frequent mutual aid drills using "dry runs" of planned hazards.

It also requires a new way of thinking. In the past, each fire department assumed the responsibility for protecting its own area of responsibility. Each engine company based its operating practices on the amount of water available from hydrants, tankers, or other accessible sources. Fire officers accepted the fact that they had to attack the fire with small lines initially, then call for assistance to sustain them.

Fire department water supply takes the approach that when no fire hydrants are readily available, the fire department has to be able to transport water from the closest usable source. This requires a coordinated effort to solve problems on a regional basis. It also mandates the purchase of equipment that will be used primarily for water supply. Automatic mutual aid with other fire departments is essential in order to attain the best fire protection at the lowest cost.

WATER SUPPLY METHODS

When the source of water is located more than 1,000 feet from the scene of the emergency, there are only two ways to get water to the source: pump it or haul it. There are four basic methods that can be used to do this:

- Direct lay from a hydrant to the scene of the incident;
- Reverse lay from the incident to the water source;
- Relay operations; and
- Water shuttles.

Each of these methods has its advantages and disadvantages.

Three problems commonly are experienced:

1. Getting the operation set up and ready to apply large quantities of water to the fire can delay operations by as much as 30 minutes or more in complex water supply layouts. When this happens, the water supply that initially would have been sufficient then is inadequate to handle the volume of fire that exists when the water begins to move and a dependable supply is assured.
2. The water supply evolution is not designed to deliver enough water for the volume of fire. Single lines of 2½" or 3" hose used to transport water will not support an attack capable of bringing a major fire under control quickly.
3. There is enough water to make an initial attack, but not enough to sustain it. Two thousand gallons of water is enough to mount an attack with two initial attack lines and one backup line, but not enough to sustain it very long. If additional water is not avail-

able within five minutes, the attack may be interrupted, allowing the fire to escalate. Once the water begins to flow, the fire department must be able to sustain it as long as it is needed.

With the use of variable flow devices, it is not possible to determine how much water is flowing at any particular time, unless the pumper is equipped with flow meters. Even flow meters do not solve the problem since the flow is under the control of each nozzle operator and can be changed at any time. If changes in the setting of the nozzle are made, the water must be there to supply it. To ensure that it will be, it is best to base water supply requirements on the maximum demand that can be placed on the system by the hoselines and water flow devices in service. It is the water supply officer's job to see that enough water will be available to meet the potential demand at the time it is needed.

DIRECT LAY FROM A HYDRANT

When operating in a hydrant area, the quickest way to get water is to have the initial engine company drop a supply line at the hydrant on the way in, and then lay hose to the fire scene. This provides a limited amount of water for the initial attack without having to wait for a second engine company to establish the water supply. This method of supply has two limitations.

First, since the only pressure available to overcome the friction loss in the hoseline is the residual pressure at the hydrant, the maximum flow is limited by the size of the hose and the pressure in the system. If the pumper is carrying 2½"-hose, the maximum flow rate will be less than 150 GPM when the distance to the hydrant is 1,000 feet or more.

Second, the available water supply can be increased by connecting a pumper in the line to pressurize the supply line. If the hose is connected directly to the hydrant, it will not be possible to make this connection without interrupting the flow to the fire, unless a four-way or hydrant gate valve has been connected to one of the unused outlets of the hydrant as shown in Figure 1-14.

Hydrant operation from a direct lay can be enhanced by using hose appliances. One way to connect a pumper into a hoseline without interrupting the flow is to use a gate valve and a clappered siamese as shown in Figure 1-10. Putting an external gate valve on an unused outlet of the hydrant before it is opened provides a means of connecting a pumper to it without closing the hydrant. If the supply line to the attack pumper is connected to the discharge side of the siamese with

16 Fire Department Water Supply Handbook

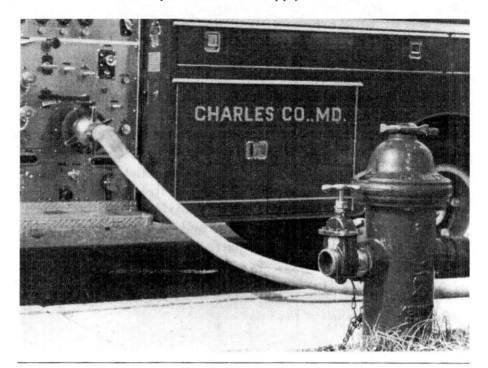

Figure 1-9. Hydrant with gate valve to provide access to unused outlet.

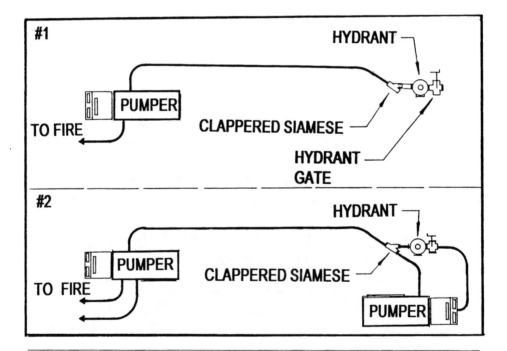

Figure 1-10. A hydrant gate valve and clappered siamese can be used to make the transition from direct supply to a relay without interupting the flow to the attack.

An Introduction to Fire Department Water Supply 17

Figure 1-11. Gate valve and siamese used to enable a pumper to be inserted into supply line.

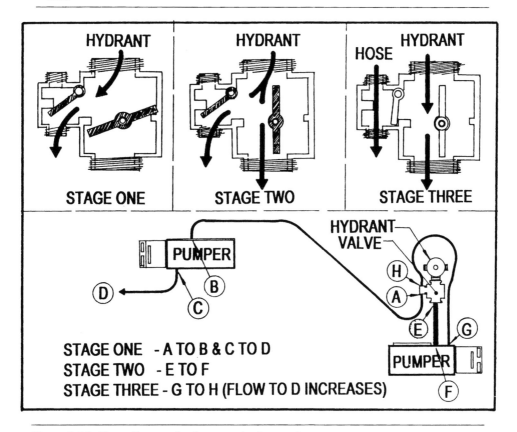

Figure 1-12. Flow diagram for four-way hydrant valve.

one of the inlets connected to the hydrant, it will be possible to insert a pumper in the supply line without interrupting the flow. Section #1 of Figure 1-10 depicts the siamese connected to the hydrant with water flowing from the hydrant through the supply line to the attack pumper. The intake to the second pumper can be connected to the gate valve on the unused outlet of the hydrant. A discharge outlet from the pumper is connected to the unused inlet of the siamese and the valve opened; the pump can be used to build up pressure after the valve is opened. When the pressure from the pumper becomes greater than the residual pressure on the hydrant, the clapper valve in the siamese will shift, allowing the second pumper to increase the pressure to the line. The increased pressure from the pump overcomes the residual pressure of the hydrant, allowing the clapper valve in the siamese to isolate the supply line from the hydrant as illustrated in Section #2 of Figure 1-10. At that point, the hoseline is supplied from the hydrant through the pumper, and the flow can be increased within the limits of the ability of the hydrant to supply it, the supply hose to the pumper, and the hoseline to move it.

Another approach to getting a pumper in the line after the supply line is in service is to use a four-way valve on the hydrant. Several different valves are available, but all of them provide a supply line outlet to the attack engine and a gated 4½"-connection and a supply line intake for the second pumper to allow it to be connected into the supply line without interrupting the flow of water.

The four-way valve is connected to the 4½"-outlet on the hydrant and the supply line is laid from the supply line outlet to the attack engine. The hydrant is opened and the valve set to allow water to flow through the supply line using the residual pressure on the water system shown as Stage 1 in Figure 1-12.

The supply engine makes the connection from the 4½"-outlet on the hydrant valve to the intake of the pump, and from one of the outlets of the pump to the supply line intake of the four-way valve.

The valve is turned to supply water to the engine through the 4½"-outlet, while still allowing water to flow through the supply line to the attack engine. This is designated as Stage 2 in Figure 1-12.

The supply engine uses its pump to provide the needed pressure for supplying water to the attack engine. The hydrant valve is set to replace the flow from the hydrant to the attack engine with water from the pressure side of the supply pumper. (Some hydrant valves are equipped with check valves which will make the change automatically when the pressure from the fire pump is higher than the residual pressure at the

An Introduction to Fire Department Water Supply

Figure 1-13. Single line of 4"-LDH supplying 375 GPM by hydrant pressure alone.

Figure 1-14. Four-way hydrant valve with soft sleeve to pumper, 3"-line to inlet of valve, and 4"-line to the attack pumper.

hydrant outlet.) Stage 3 in Figure 1-12 provides full flow through the system.

The use of Large Diameter Hose (LDH) reduces the need for these methods where there is adequate hydrant pressure by helping overcome the limitations imposed by the friction loss of fire hose. Table 1.2 estimates the distance water could be moved through a single line of hose, connected to a hydrant with 50 PSI residual pressure, using hoselines of different sizes.

Table 1.2

Hose Diameter	500 GPM	1,000 GPM	2,000 GPM
2½"	100'	—	—
3"	200'	—	—
4"	1,000'	250'	—
5"	3,000'	800'	200'
6"	6,000'	1,500'	400'

Table 1.3

Hose Diameter	500 GPM	1,000 GPM	2,000 GPM
2½"	—	—	—
3"	100'	—	—
4"	400'	100'	—
5"	1,250'	350'	—
6"	2,400'	600'	150'

Using a direct lay from the hydrant will provide an initial supply of water in a relatively short time, but it would be difficult to meet the eight-minute objective of sustaining a 400-GPM attack with an initial supply of 2,000 gallons on board the first alarm units. Table 1.2 also shows that the maximum distance from the hydrant to the fire would be limited when using anything smaller than 5"-hose. The maximum distance also will be reduced accordingly if the residual pressure at the hydrant falls below 50 PSI. Table 1.3 shows the maximum distance if the residual pressure were reduced to 20 PSI, the minimum acceptable pressure at the hydrant.

Four-way hydrant valves are available that can be used with LDH and have 4"- or 5"-inlet and outlet fittings instead of 2½"-couplings. LDH four-way hydrant valves are used primarily when fire hydrants can supply large flows with limited residual pressure.

REVERSE LAY FROM THE SOURCE

When the standard response includes two engine companies, the first-in engine company may be dedicated to making an initial attack using the water in its tank. The second-arriving engine company drops off extra firefighters to assist with the attack, then proceeds to lay a line to the source to provide the water supply. This approach works well when both engine companies are responding simultaneously, and will arrive on the scene at nearly the same time. A problem arises when the second engine company is delayed, and the water carried in the tank of the attack engine is exhausted before the supply line can be put into service.

In rural areas, the closest water supply may be a pond or stream. To use it requires a pumper at draft. In this situation, one approach is for the first-arriving engine company to lay a supply line from the scene to the water source. The same pumper can then set up a draft and begin to supply water to the scene through this hoseline. This type of operation, however, has many disadvantages.

The most obvious disadvantage is that the pumper is not located at the scene of the fire, but is set up to draft some distance away. When a reverse lay is standard operating procedure for an engine company,

Figure 1-15. A reverse lay of 2½"-hose from fire to a farm pond.

Figure 1-16. A manifold can be used to supply a number of lines from a reverse lay of LDH.

arrangements need to be made in advance to leave the required hose and equipment at the fire scene before beginning to lay hose to the water source. Skid loads can be used to provide the needed attack lines on the scene, ready for use when the supply lines are charged. The use of a manifold or "phantom hydrant" at the fire scene offers some flexibility for making changes and controlling flow; it also enables a second pumper to be connected into the line when it becomes available. Kits of special equipment can be prepared to drop off before the pumper leaves the scene to make the reverse hose lay. Each firefighter should be responsible for obtaining certain equipment that will be needed for the attack, including breathing apparatus and spare bottles, before the pumper leaves to go to the water source. Even when standard procedures address the inconveniences inherent in this type of operation, firefighting operations are likely to be difficult. Every effort should be made to dispatch enough apparatus to keep a pumper on the scene to supply the attack lines.

Reverse lays often provide an inadequate amount of water to overcome the fire. Because of the distances involved and the limited amount of hose on a pumper, a single supply line frequently is used. Large diameter hose is one solution to this problem. Using 5" or larger hose with a manifold at the fire scene, can provide 1,000 GPM or more to supply a number of attack lines up to a distance of 2,000 feet from the water source. Figure 1-16 shows a manifold being supplied by LDH with a number of other lines to the fire.

As additional engine companies arrive, a reverse lay to the source can be converted into a relay. The supply pumper at the source can use additional hoselines, laid to an attack pumper on the scene, to supplement or replace the limited attack capability that was initially available.

When a single engine company responds to a fire in a rural area, a reverse lay to the source may be the only practical way to get water to the fire. The biggest disadvantage in using the first-arriving engine company in this manner is that the attack on the fire cannot begin until the water supply is in place and operational. With other methods of operation, the initial engine company makes the attack using transported water, while the second engine company furnishes the water supply. In this (reverse) case, the engine that arrived on the scene first is engaged in setting up the supply, and the attack has to wait until another engine company arrives or the water supply has been established. The need for water supply is one good reason why single-engine-company response to building fires is not desirable. If only one engine company is available for response at certain times, arrangements should be made with a neighboring department to supplement the response with automatic aid.

RELAY OPERATION

Historically, the standard approach to moving water over long distances has been by relay. Pumpers in the line at intervals are used to overcome the pressure losses experienced as the water moves through the hose. Two problems limit the effectiveness of relay operations.

First, it takes too long to set up and begin the relay operation; second, the completed layout may be unable to supply enough water to meet the needs of the situation.

Recent developments in fire service technology have brought improvements in both of these areas.

High-capacity pumpers equipped with intake relief valves are readily available in most areas. The prevalence of 1,000 GPM or larger pumpers minimizes the likelihood of a small pumper at the source of a relay limiting the total flow. The use of intake relief valves on the terminal end of a relay prevents pressure surges or uncontrollably high residual pressure at the attack pumper when less water is being used, and the friction loss in the supply line decreases.

Large diameter hose has the potential to greatly increase fire flows. A 4"-supply line will provide 750 GPM, up to 1,000 feet, without a relay pumper. Five or 6"-hose increases the amount of water that can be supplied even more, and reduces the number of pumpers needed to complete the relay. As an example, one pumper can supply 1,000 GPM through nearly a mile of 6"-hose. The ability to limit the number of pumpers required in the relay not only reduces the delay in setting it up, but makes the operation much simpler and smoother while it is flowing water.

A relay operation is most effective when large amounts of water are required over a long period of time. Once a relay is established, it can function with very little adjustment and a minimum of changes. At the same time, the delay in getting started precludes meeting the eight-minute objective in most cases. Chapter 8 of this book, Relay Operations, covers relay operations in detail.

WATER SHUTTLE OPERATION

Using tankers to haul water to the fire scene is becoming more common, even in fire departments that operate in suburban and urban areas with good water systems. Using a water shuttle to provide the needed fire flow offers many advantages.

Using portable reservoirs or nurse tankers to maintain a reserve water supply on the fire scene while other units are shuttling water can maintain fire flows as high as 2,500 GPM for extended periods of time. High-capacity water shuttles with a continuous flow capability in excess of 500 GPM have the potential to:

- Extend fire protection capabilities beyond the areas that are served by water systems and make the type of attack on the fire that is required to accomplish rapid extinguishment;
- Provide enough water to use master streams and large attack lines effectively;
- Maintain the rate of flow needed to protect exposures, control fires involving large amounts of flammable materials, and handle emergencies involving dangerous substances;
- Take advantage of water sources, either hydrant or static, that are remote from the scene of the emergency;
- Provide the needed water supply for HAZMAT teams to function safely; and
- Overcome the limitations imposed by reduced manning in volunteer departments.

An Introduction to Fire Department Water Supply

Figure 1-17. A portable tank can provide on-site storage to sustain a water shuttle.

Figure 1-18. High-capacity water shuttle supplying 2,850 GPM.

Water shuttles provide a more dependable supply of water than any other method because:

- A number of tankers are involved in a water shuttle and a mechanical failure will only cause a reduction in water supply, not a complete interruption, which might be the case when using a single line for direct supply from a hydrant, a relay operation, or a single-tanker response to an incident; and
- If problems are experienced at the primary fill site, it is simple to divert tankers to an alternate location for filling.

Since a water shuttle's capacity is limited by the tankers that are being used to haul water, auxiliary equipment such as transit-mix concrete trucks, tankers used for filling swimming pools, or other types of vehicles can easily be used to supplement the fire department apparatus when more water is needed. Very few modifications are needed to use these types of vehicles, but advance planning, training, and arrangements for dispatching are essential.

Probably the biggest advantage to using a water shuttle is the time required to get it organized and begin delivering water. If a shuttle is organized properly, tankers should be arriving at the fire scene on a regular basis soon after arrival of the first engine company, and should be able to maintain a continuous supply from that point. There is no question that the eight-minute objective can be met if high hazard structures are preplanned with an eye toward providing an adequate water supply for the degree of risk that is present.

ISO RATING

In the 1980 issue of its Fire Suppression Rating Schedule, the ISO made provisions for fire department water supply as an alternate means of providing the required fire flow where hydrants are not located within 1,000 feet of a structure. To receive credit under this schedule, the fire department must demonstrate that it is able to attain a fire flow of at least 250 GPM within five minutes of arrival and to maintain it without interruption for at least two hours. This can be done by pumping it, hauling it, or any combination of the two. With Fire Department Water Supply, some areas have been rated as high as Class 5, using only tankers, with no hydrants located within their territory.

NFPA STANDARDS

In 1231, *Standards on Water Supplies for Suburban and Rural Fire Fighting*, the NFPA has set some guidelines for needed fire flow in areas that are not served by public water systems. This document both sets standards and provides a wealth of information about how to meet them.

THE JOB OF THE WATER SUPPLY OFFICER

The technology exists today to supply enough water to almost any location for effective fire suppression efforts. Unfortunately, some very efficient and well-managed fire departments cannot supply it where fire hydrants are not readily available. Fire departments must become proficient in supplying water to the fireground when it is needed if they are to provide the protection that is expected by all of their citizens, not just those who happen to live in areas served by a central water supply system with hydrants readily available to the fire department.

The key to all of this is establishing an organization within the fire department to ensure that all water supply needs can be met. The water supply officer is the key player. This book can be used as a detailed and practical guide for setting up an adequate water supply under a variety of situations.

2. Water Supply Master Planning

AMONG THE MANY APPROACHES to fire protection planning, most rely heavily on statistics, opinion surveys, and historical data. However, the end product of this process often is a continuation of past practices. This means that serious deficiencies may not be corrected, and that poor operating practices not only continue, but worsen. Using water supply as a basis for fire protection planning provides a more realistic view of all the problems, and establishes a reference for improvement. Water supply planning uses an objective, instead of a subjective, approach to improving the level of fire protection in a community.

Fire protection planning, based on water supply, is a four-step process.

Planning begins with an in-depth evaluation of the level of risk in the fire management area (FMA) or areas the department serves. This includes determining the required fire flows in the FMAs.

After the risk evaluation, an accurate estimate is prepared of the department's ability to deliver the needed fire flow to meet identified needs.

Next, you determine how to get the most out of your presently available resources; this is the most cost-effective way to improve the level of protection.

Finally, you develop a long-range plan for improvement. This plan should deal with the deficiencies that have been identified. Ideally, this brings the level of unprotected risk to a point that is acceptable to the community.

GUIDELINES FOR PLANNING

The first place many fire departments look when beginning a comprehensive planning process is the Insurance Services Office (ISO). ISO's rating guide is a helpful reference since it provides a list of areas of concern that the evaluator will study when assigning a public fire protection classification. Since the insurance rates in a community may be determined by the ISO's rating of the fire defenses, it makes sense to consider its recommendations. When insurance premiums are reduced, the benefits of a well-equipped fire department are graphically illustrated for the department's customers. The potential savings to taxpayers in the form of reduced insurance premiums can give the governing body the incentive it needs to provide the funds necessary for a fire department to operate efficiently.

Years of experience have given the ISO a wealth of historical data to use as a basis for its rating methods of both the fire department and the water system. At the same time, the ISO tends to look backwards instead of to the future; that is, its recommendations generally rely on tried and true methods of operation; many of the latest advances in fire protection technology are not included in the current issue of the rating guide.

The National Fire Protection Association (NFPA) publishes several more current standards that provide a better basis for improvement. NFPA 1231, *Standard on Water Supplies for Surburban and Rural Fire Fighting*, is particularly helpful. It not only suggests methods for determining fire risk and estimating the needed fire flow, but details how fire flow can be supplied. This standard should be used as a guide throughout your planning process. Many of the methods and procedures suggested in this document can help fire departments in urban areas, even in some of the largest cities.

Another publication that should be considered is NFPA 1500, *Standard on Fire Department Occupational Safety and Health Program*, which covers many aspects of fire department operations, including recommended standards for personnel, training, apparatus, equipment, and operating practices and procedures, in addition to safety in general. Adhering to the NFPA 1500 recommendations will improve a department's efficiency as well as reduce dangers inherent in fire suppression activities. Another concern is the legal liability of both the fire department and the local authority having jurisdiction; both are responsible for maintaining a safe environment for all fire department personnel.

THE PLANNING PROCESS

Effective fire protection planning requires that fire departments, local governments, or both, establish an orderly process to ensure that all aspects of fire department operations, including water supply, are included. One approach that has worked well is shown in the flow chart, Figure 2-1.

The process begins with a risk analysis of the area of responsibility, determining both general fire flow requirements and those for specific target hazards. Coverage of the area in terms of fire stations and apparatus, water distribution systems, and other water sources, is plotted on a detailed map. Fire department capabilities are measured by conducting an inventory of equipment, apparatus, and personnel responding to alarms. Fire apparatus is evaluated, and the flow capabilities of any water systems are verified where needed.

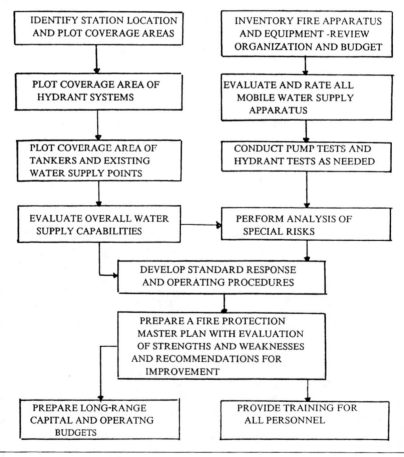

Figure 2-1. Flow chart for creating a fire protection master plan.

Based on all the collected data, the fire department's overall ability to meet the needs of its area of responsibility is evaluated. Recommendations for improvement are made, including long-range planning for additional fire stations, apparatus, equipment, and personnel.

A standard response to alarms and standard operating practices and procedures are adopted to meet the needs of the area being served. These standards should be based on both existing fire suppression capabilities and immediate necessary changes, as well as incorporating expected changes that will be made as the long-range plan is implemented. The resources of nearby departments (personnel, apparatus, and equipment) should be considered if they are part of a formal mutual aid agreement.

As part of the planning process, the department's training program is evaluated and plans for improvements made. To help firefighters retain their job skills, an organized in-service training program, in which each firefighter participates on a regular basis, is essential. Specialized training programs for officers, pump operators, and firefighters, which concentrate on teaching water transport skills, should be offered and all officers or prospective officers required to participate in them.

Intercompany drills and training sessions conducted on an area-wide basis are also important. Using fire department water supply techniques requires that a number of different departments work as a team to get the most out of the apparatus, personnel, and equipment (APEs) that each can contribute. Unless the departments are accustomed to working together, and unless they use similar procedures, they will not be very effective in an emergency situation. Joint training sessions also are critical during an ISO evaluation. ISO allows a fire department to claim credit for the resources that are available from departments within a five-mile radius; however, the amount of credit received depends on the "Automatic Aid" (AA) factor. Automatic Aid requires that the assisting department(s) be dispatched automatically on all structure fires in the area being rated. Under the provisions of a written agreement, automatic aid units must be dispatched at the same time as the first-due companies. Once the basic requirement has been met for station location and automatic dispatch, the AA factor is applied to the APEs that will be responding. The largest single factor in determining how much credit will be allowed is the number of inter-department training sessions that have been conducted. To receive maximum credit, all of the departments involved in the automatic aid agreement must participate in at least four half-day sessions annually.

RISK ANALYSIS

Fire flow requirements are based on the number and type of structures in the fire department's area of responsibility, and on any special hazards such as hazardous materials storage, large commercial buildings, and schools. Serious life hazards are present in hospitals, nursing homes, prisons and critical care facilities, and create special problems for the fire department. Large buildings require fire flows well in excess of 1,000 GPM. There are a number of ways to evaluate the level of risk. The National Fire Academy has developed a method of risk analysis that is used in its Community Fire Defense curriculum. It is very detailed, and can be used to accurately calculate needed fire flow. ISO recommends a method of risk analysis in its rating guide that also provides a good estimate of the potential fire flow requirements for a variety of structures and occupancies. Using the ISO method has one major advantage: it is the one an ISO evaluator will use to rate your department's required fire flow and number of engine companies required. Both of these methods are rather cumbersome; some fire officers may not feel comfortable using them.

As an alternative to these two approaches, NFPA 1231 specifies a method of risk analysis that considers most of the significant factors in estimating required fire flow, but is simpler and easier to use. If NFPA 1231 is used in other parts of the planning process, using it as a guide for risk analysis should be more compatible with the rest of the plan and therefore more convenient to use. Chapter 3, Risk Analysis in Fire Protection Planning, covers risk analysis in more detail.

Any one of these methods of risk analysis seems to indicate that a minimum fire flow of 500 GPM would be required to protect areas that are residential in nature. According to NFPA 1231, single-family dwellings require a fire flow of at least 500 GPM. The ISO rating guide specifies that single-family detached dwellings, located a minimum of 100 feet apart, need a fire flow of 500 GPM. Where single-family detached dwellings are closer together, or when larger buildings are involved, a larger base fire flow will be needed to handle the added risk.

When the fire department's entire area of responsibility is similar in terms of hazards, construction, etc., an overall base fire flow can be established, while target hazards and large structures are dealt with on an individual basis. Where there are clusters of structures that require more than 500 GPM, fire management areas can be established. Planning fire protection for individual FMAs is based on the required fire flow within each of them.

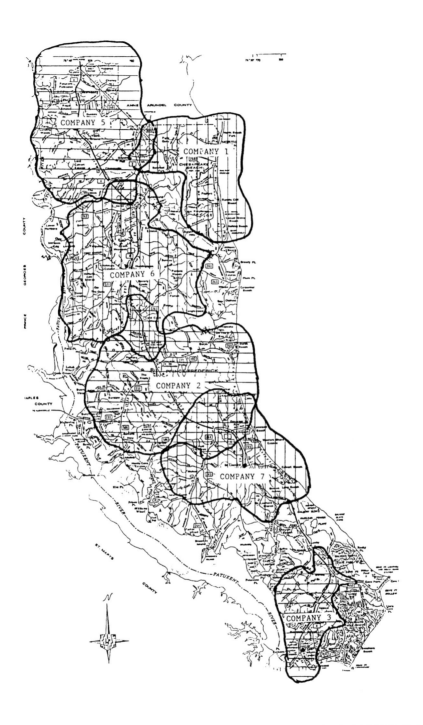

Figure 2-2. Five-mile coverage map for fire stations in Calvert County, MD.

FIRE STATION COVERAGE

The ISO considers any structure that is more than five miles from the closest fire station to have inadequate fire protection. These structures are rated Class 10, and the highest fire insurance premium rates are applied. From a life and property protection standpoint, it is very difficult for a fire department to be effective when initial response companies have to travel more than five miles. Longer travel times mean increased fire spread before the first engine company arrives.

The five-mile coverage criterion is only a minimum standard. As part of its evaluations, the ISO estimates the percentage of the built-up area being protected that is within one and a half miles of the nearest engine company, and within two and a half miles of the nearest ladder or service company. The percentage is calculated by comparing the number of fire hydrants within the required distance to the number that are not. For adequate fire protection, a good objective would be to have an engine company within one and a half miles of any high value fire management area, a ladder or service company within two and a half miles, and no structure more than five miles from the closest fire station.

A detailed coverage map should be prepared which shows the location of each fire station and its five-mile coverage area, based on road mileage. Keep in mind that coverage maps based on a five-mile radius from each fire station can be deceptive because locations that appear to be geographically very close to a fire station when the distance is measured in a straight line could actually be much further away by road, because of rivers, mountain ranges, or other physical barriers.

Figure 2-2 is an example of a fire station coverage map. In this example, you can see an area in the southern end of the county that is more than five miles from any fire station. One of the large housing developments located within the unprotected area has a central water system and presently enjoys a Class 7 ISO rating. At the time the community was last evaluated, one of the neighboring fire departments had a pumper housed in a station that was located within the boundaries of the development. This station is no longer in service. It would appear that when ISO re-evaluates this area, the PPC rating will change from a Class 7 to a Class 10. This would increase fire insurance premiums paid by property owners approximately $100,000.00 per year unless a fire station is added to serve this community. It is only when the fire protection problem is surveyed on a comprehensive basis that this type of deficiency is likely to be found.

HYDRANT SYSTEM COVERAGE

The most dependable water supply is a well-designed and maintained public water system equipped with properly spaced fire hydrants. To attain the maximum credit of 1,000 GPM from an individual hydrant during an ISO evaluation, it must be within 300 feet of the structure that is being rated. A maximum of 670 GPM can be credited for a hydrant within 600 feet of the test location, and 250 GPM for hydrants between 601 and 1,000 feet. This ISO limitation means that many buildings within the service area of a water system are considered by ISO to be underprotected because of hydrant spacing. When this is the case, the fire department must be able to supplement the flow from the closest hydrant by transporting water over longer distances to obtain the maximum flow from the system.

By using large diameter hose (LDH) and tanker shuttles, many fire departments can move large volumes of water over long distances. Under these conditions, the hydrant system coverage map can be extended as far as the fire department can transport the needed fire flow. As an example, 5,000 feet of 5"-hose would extend a fire department's coverage map to 5,000 feet from the last hydrant in the system instead of 1,000 feet. If the department is capable of transporting 250 GPM within 5 minutes of arrival and the desired fire flow by tanker shuttle over a distance of 2 miles within 15 minutes, the coverage map can be extended 2 miles from the last hydrant.

When water systems are old, have limited storage, or have inadequate or poorly maintained distribution systems, they are not capable of supplying the needed fire flow. A supplemental supply will be required; this should be noted on the coverage map. Plans to supplement the system should be included in your response SOPs.

TANKER COVERAGE

Mobile water supply apparatus are required to furnish the needed water supply in areas that are beyond the reach of water systems, suction supply points, and hoselines. To sustain an initial attack, tankers need to be distributed throughout the response area so they can be dispatched to any emergency before the initial engine company's water supply is exhausted.

While travel time is more important than actual distance, the five-mile coverage map can be used as a guide. If you are attempting to qual-

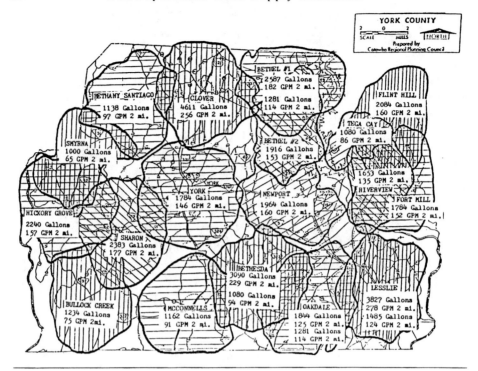

Figure 2-3. Five-mile coverage map for existing tankers in York County, SC.

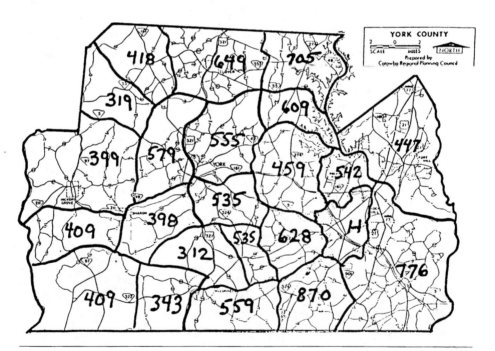

Figure 2-4. Constant flow capabilities by tanker response areas in York County, SC using existing tankers.

Water Supply Master Planning

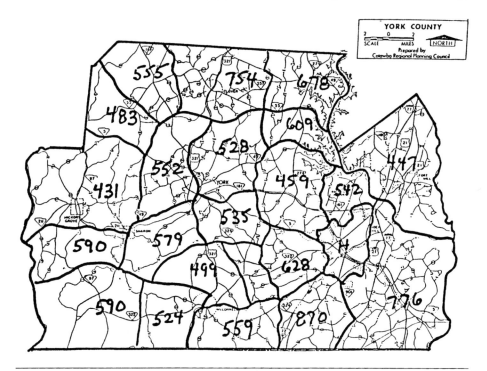

Figure 2-5. Constant flow capabilities by tanker response areas in York County, SC after two of seventeen tankers have been replaced.

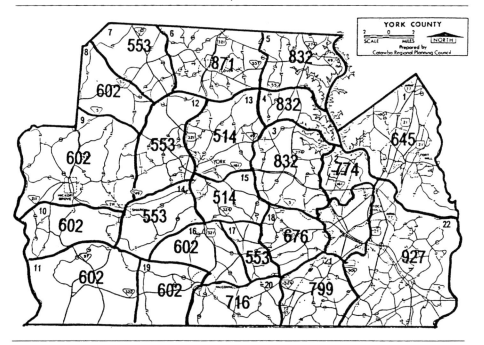

Figure 2-6. Constant flow capabilities by tanker response areas in York County, SC at the completion of the tanker replacement program five years later.

ify for an ISO rating lower than Class 9 and you are using tankers to haul water into areas that are not protected by fire hydrants, remember: there is a limit to the distance mutual aid tankers are allowed to travel. As a general rule, ISO will allow credit only for tankers that are housed within five road miles of the boundaries of the coverage area of the department being evaluated. This five-mile coverage map (Figure 2-3) provides a basis for planning to meet ISO's requirements.

Stations that house a vehicle that can be described as a mobile water supply apparatus should be noted on the coverage map. By including the amount of usable water each tanker carries, and the GPM rating assigned to it, the estimated fire flow in any portion of the coverage area can be determined. Figure 2-3 is one example of such a map; the plan is based on using tankers to haul water as far as two miles from the source. In this example, each of the 17 fire stations operates one tanker, and several of them have two. Figure 2-4 shows the available fire flow from these tankers within response areas, assuming three departments are dispatched on all structure alarms. As part of the evaluation and comprehensive plans, recommendations were made for improving some of the tankers, and replacing others. Of the 22 tankers in service in this county, it was recommended that two of the least efficient ones be replaced. Figure 2-5 shows the same response areas, but estimates the available fire flow after replacing the two inefficient tankers and making minor modifications to the others. In this case, replacing 2 of the 22 tankers (9 percent of the total) resulted in an improvment in the average fire flow in the western third of the county from 398 GPM to 533 GPM, an increase of nearly 34% in the available flow. Figure 2-6 shows the expected flow in the same areas five years later, after the completion of the tanker replacement program in York county.

SUCTION SUPPLY POINT COVERAGE

When fire hydrants are not available, suction supply points can be used to get water for direct application to a fire, supplying a relay, or for filling tankers. A suction supply point is a static source of water that has been identified as accessible to a pumper for drafting. To be included in a water supply plan, it should have all-weather accessibility; in addition, the fire department must have written permission from the owner to use it when needed. Except in special cases (such as where a storage tank was used as a water source), a draft supply point should be able to sustain the full rated capacity of any of the pumpers that might use it for at least two hours of continuous pumping. To meet the two-hour pumping criterion, the size of the body of water should provide enough storage, the replenishment rate of the pond or lake should be high enough, or the flow rate of a stream or river must be sufficient. Chapter 7, Suction Supply as a Water Source, covers suction supply in detail.

A coverage pattern should be developed for each suction supply point. The size of these coverage patterns will be limited by the department's ability to move the needed fire flow using fire hose or tanker shuttles. Once the maximum practical distance the water can be transported has been determined, the coverage pattern for each suction supply point, as well as the coverage patterns for any available water systems, should be inscribed on a detailed map and areas of inadequate water supply identified. Figure 2-6 is an example of a water supply point coverage map based on the ability of the tankers in the area to haul water two miles from the fill point.

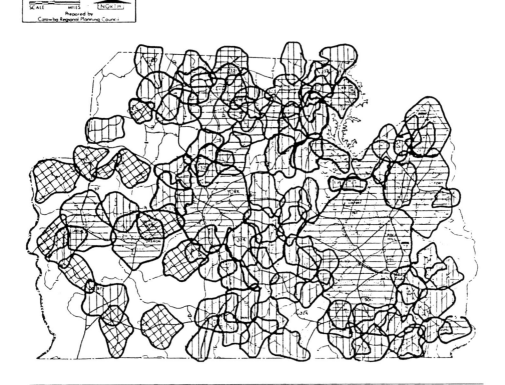

Figure 2-7. Two-mile coverage map of water supply points in York County, SC.

PERSONNEL

To evaluate whether your response staffing levels are adequate, you need a record of the number and types of alarms for the preceding three years, along with the average number of personnel responding to each of them. An analysis of the types of risk and the required fire flow within the department's area of responsibility should determine how many firefighters would be needed to handle a typical alarm.

An inventory of the department's apparatus and equipment can be related to the needed fire flow. If there are not enough personnel to operate additional apparatus and equipment effectively, it is pointless to purchase any. If there is a shortage of personnel, consider such alternatives as a volunteer recruiting program, on-call paid personnel, additional full-time paid personnel on duty, or some combination of these.

If adding personnel is not an option, specialized equipment and large hoselines enable a department to apply the maximum amount of water to a fire with a minimum crew. The use of 1¾"- or 2"-attack lines, equipped with the new low-pressure nozzles, enables two or three firefighters to apply more water, with more flexibility, than the traditional 1½"- or 2½"-attack lines. Once it is set up, a master stream device can be used to apply up to 1,000 GPM to a fire; only one firefighter is needed to direct the stream.

A single line of large diameter hose has the potential to supply the full capacity of a pumper over distances greater than 1,000 feet. In many systems, large diameter hose can eliminate the need for a pumper at the hydrant, allowing the department to respond with one less pumper to an incident, and freeing the firefighter who would have driven it to assist in the fire attack. One 1,500-GPM pumper can take the place of two 750-GPM pumpers in some applications, again releasing a firefighter from the responsibility of operating the additional pumper.

APPARATUS

Inventory and review the apparatus normally housed in each station. Keep in mind that the amount and type of apparatus needed depend on the area to be protected. The ISO requires one engine company for fire flows up to 1,000 GPM, two companies for a basic flow of 1,250 to 2,500 GPM, and three in areas with a required fire flow of 3,000 to 3,500 GPM. These engine companies can respond from sev-

Figure 2-8. Fully equipped rural fire station in Calvert County, MD-2 Pumpers, a Pumper-tanker, a Rescue truck, an Ambulance, and a Brush Truck.

eral different stations, all within the coverage area. Any of these requirements can be satisfied in part by an automatic aid system.

In general, each station should have at least one engine company equipped with a standard pumper, with a second pumper in reserve. The reserve pumper can be an older model, but it must be well maintained and equipped, and reliable.

Each station that protects an area with structures located more than 1,000 feet from the closest water source should have some type of mobile water supply apparatus in addition to an attack pumper. A pumper tanker or general purpose tanker can be used for water supply, with a secondary role as a reserve engine. With this approach, keep in mind that the water supply apparatus would not be available for its intended purpose when it is being used for fire attack by the engine company. Another possibility is for a number of stations to share a reserve pumper. The 1980 issue of the ISO rating guide allows each eight engine companies to share a reserve pumper; however, one in each station would provide a more dependable response.

The 1980 issue of the ISO rating guide also suggests that a ladder company or service company is required within two and a half miles of the built-up area that is being protected. You need a ladder company when there are more than five buildings of three or more stories (35 feet in height or more), your needed fire flow is 3,500 GPM or greater, or both. When a ladder company is not needed, a utility or squad type

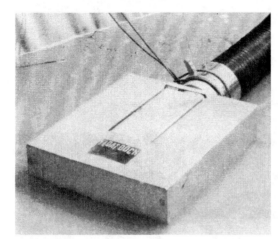

Figure 2-9. Floating strainer.

Figure 2-10. Low-level strainer.

Figure 2-11. Automatic Suction Valve.

vehicle can carry the equipment that a service company needs to do the job. This can be a multi-purpose vehicle, equipped with a booster pump and a small water tank, that can handle miscellaneous alarms without committing a full engine company.

Small departments may not have a dedicated service truck, but service type equipment still will be needed. ISO gives partial credit for a combined engine-service company, but does not allow full credit. Be aware that adding service type equipment to a fully equipped pumper may overload it.

Good long-range planning requires estimating the remaining useful life of each piece of apparatus, and developing a long-range plan for replacement when it is most cost effective to do so. In a multi-station environment, it is a good practice to rotate apparatus among stations of varying activity levels to equalize the wear and tear on the vehicles, to avoid replacing them prematurely. As a vehicle approaches the end of its useful life, convert it to reserve apparatus.

An apparatus replacement schedule is an essential component of a long-range capital budget. Too frequently fire departments find themselves needing to replace a number of vehicles at the same time, with no funding available to do so. Long-range planning enables a fire department to set up and regularly contribute to an apparatus replacement fund to use for unexpected expenditures. It also helps the department avoid expensive financing arrangements.

EQUIPMENT

The ISO lists the equipment that should be considered standard for engine, ladder, and service companies. When no ladder or service company is available, the evaluator will allow partial credit for the additional equipment if it is carried on a pumper. Considering the required engine company equipment, and other items that are needed for fire department water supply, a separate service truck is more useful.

The minimum standard for equipment recommended by the ISO may be adequate for a fire department that has all of the personnel it needs, and an efficient water system. Most fire departments will need additional equipment both for water supply and to furnish a satisfactory level of fire protection.

Large diameter hose, along with the necessary fittings and adapters to use it efficiently, is one of the most cost-effective purchases a fire department can make. With a single line of 5"-hose, one pumper can deliver as much as 1,000 GPM over a distance of 2,000 feet without a

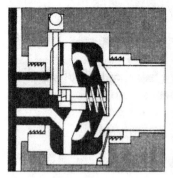 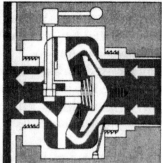 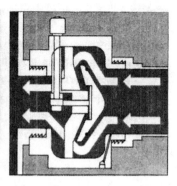

VALVE CLOSED
Automatically shuts off when water flow stops (can be manually locked closed). Valve will automatically open when flow is resumed.

VALVE OPEN (Free Floating)
Automatically adjusts to changes in flow and pressure. Automatically closes to reduce water hammer if water flow is abruptly stopped.

VALVE FULLY OPEN
Low friction, streamlined design for maximum flow (can be manually locked open). Low pressure area behind float enhances laminar flow.

Figure 2-12. Operating diagram of automatic suction valve.

relay pumper in the line. Using 1¾''- and 2''-hose as attack lines instead of 1½''-hose also can deliver more water with less effort. While none of these is listed as required in the 1980 issue of the ISO rating guide, they are necessary if the fire department is to maximize the engine companies that are available.

A floating strainer is essential when the department uses suction supply points for water supply. With a floating strainer a pumper can deliver its full capacity from water sources as shallow as 18 inches deep. When mobile water supply apparatus is available, a portable tank is needed so the tanker can empty its load of water quickly at the fire scene for the attack engine to use. The tanker then can reload water while the attack pumper drafts from the portable tank to supply the attack lines. The drafting pumper must have either a low-level or floating strainer to draw down the level in the tank as low as possible before it loses its prime. A manual shut-off valve on the large suction inlet connects the suction hose to the attack engine without interrupting the initial attack being made from the tank on the apparatus. There is an automatic suction valve that can be used on the suction inlet; it enables the attack engine to make the transition from the tank on the apparatus to the portable tank after it has been filled without losing the prime on the pump. Figure 2-12 illustrates an automatic suction valve. This type of specialized equipment should be included on the inventory, along with the ISO requirements.

APPARATUS EVALUATION

The condition and efficiency of a department's apparatus are more important than its age. As an example, in evaluating engine company equipment and efficiency, the ISO puts as much emphasis on the annual service tests for pumpers as it does on any other single item. According to ISO's 1980 rating guide, 100 points of a maximum 654 possible points depend on a record of annual pump service tests. Another 50 points depend on whether annual hose tests are conducted. Regular service tests ensure that a pumper can deliver its rated capacity when it is needed, and enable the department to schedule major maintenance on the fire pump. Any fire department evaluation should include verification that these tests are conducted on a regular basis, and that detailed records are available for inspection at any time. To receive maximum credit from ISO field representatives, a department's test records must be available for the preceding three years.

An in-depth evaluation of all mobile water supply apparatus is essential to estimate a department's transport capabilities with a water shuttle. The primary objective of tanker testing is to determine the amount of usable water each one carries, the loading time, and the unloading time. With this information it is possible to calculate a Continuous Flow Capability (CFC) rating in GPM for each tanker based on its capacity, the handling time, and the travel time for any given area or specific risk. A secondary objective is to determine how to increase the amount of usable water a tanker carries, or reduce the time required to load and unload it.

COST EFFECTIVENESS

It generally is not cost effective to furnish more apparatus and equipment than a fire department has the personnel to operate. In many fire stations there are as many as five pieces of apparatus, but only an average response of fewer than ten firefighters per alarm. Equipment, other than reserve apparatus, that is left behind, with no personnel to operate it, serves little useful purpose and could be considered a poor use of resources.

An exception is where specialized apparatus or equipment is required to meet specific situations. While incident commanders like to choose which apparatus and equipment will respond, it generally is more cost effective to house specialized apparatus where there are enough personnel to operate it when it is needed.

WATER SYSTEM EVALUATION

There should be a regular program of hydrant testing and maintenance. If you have these test records, you can evaluate whether the system can deliver the required flow for any fire management area or specific risk and whether supplemental supplies are needed. Without records, you may have to conduct flow tests in key locations. You must know the maximum flow rate if a hydrant is to be used as a water fill point for tankers when a water shuttle is used to supply a fire management area the hydrant system does not serve.

OVERALL WATER SUPPLY COVERAGE

An overall water supply coverage map should be prepared based on the hydrant system, draft supply points, and the fire department's ability to transport water. When the hydrant system or suction supply points cannot provide the required fire flow for a FMA, the fire hose that is available cannot move it, or both, a water shuttle should be considered as a means of fire department water supply.

The results of your tanker testing can be analyzed and used to determine the continuous flow capability of each tanker. With this information, you should be able to determine how many tankers would provide the required fire flow in any inadequately protected areas. Analyzing the expected maximum flow based on an area-wide study enables the fire department to get the most protection for the least money.

Once a department's water transport capabilities have been estimated, an overall water supply map that combines the coverage areas of the water system and suction supply points, should be prepared. Be certain to identify: areas not within range of any existing water source; potential water supply points that are not accessible to a supply pumper; locked gates; and any other problems. Those in strategic locations that can fill any gaps should be developed as suction supply points for tankers. This development could require constructing access roads, dredging a pond or stream for more depth, or installing a dry hydrant or some other means of drafting. In some cases, it will be necessary to bury a storage tank or build a cistern to provide a water supply in areas where no other option is available. Once these needed improvements have been prioritized, schedule their completion and determine how to fund them. Figure 2-7 depicts the area that will be

Figure 2-13. A ladder pipe and deck gun being supplied from a water shuttle.

within two miles of a usable water source in one county in South Carolina after all of the needed improvements have been made.

PROTECTING TARGET HAZARDS

After you have calculated the overall fire flow rating of the entire area, a risk analysis and preplan should be completed or revised for selected target hazards. The required fire flow for specific structures should be determined based on their size, occupancy, type of construction, and exposures. If more than one source is available, evaluate each to determine the best method of supplying the required fire flow. Where a water shuttle is necessary, alternate fill sites should be included in the prefire plan, along with an estimate of the number of tankers needed. Consider using the plan that was developed as part of the individual risk analysis procedure to conduct a "dry run" on one of these hazards.

FIRE PROTECTION MASTER PLAN

After all of the coverage maps have been completed and specific risk analyses done, evaluate the level of fire protection presently available in the survey area. After you identify deficiencies, and determine the level of unprotected risk, develop a fire protection master plan that will satisfy the needs of the community. The long-range plan should bring unprotected risk to a level that is acceptable to the fire department, the local governing body, and the taxpayers. The plan should project the fire protection needs for at least five years, with interim goals and objectives established within the funding levels that are available.

STANDARD RESPONSE AND OPERATING PROCEDURES

Based on the fire department's evaluation, and the identified deficiencies, standard responses to alarms and standard operating procedures (SOPs) should be adopted. The SOPs should be comprehensive enough to best use all the resources available, including mutual aid, yet flexible enough to incorporate the master plan's changes as they are implemented.

PERSONNEL TRAINING

One of the ISO's primary concerns in evaluating a fire department is personnel training. The ISO reviews records of training sessions that have been held, what material was covered, and which individual members participated. Individual firefighter training records should show in detail what subjects were covered, what skills were reviewed, and how many hours each person spent in training. These training records are not only a source of documentation for the ISO, but are used to meet the requirements of NFPA 1500, *Standard on Fire Department Occupational Safety and Health Program*. The training records are essential for protection if a lawsuit is filed against the fire department and its leadership that contends that a particular firefighter's training was not adequate, and resulted in personal injury or property loss.

While many firefighters have been participating in the certification process, and have met the requirements of NFPA 1001, *Standard for Fire Fighter Professional Qualifications*, many departments have no organized program to ensure that individual firefighters retain their skills. Each department should have a training officer, preferably one who is qualified as an instructor through his or her state's certification board. This officer should prepare and implement an organized program that will, over a certain period of time, ensure that each member of the department has an opportunity to review each of the skills that was required for firefighter certification.

It also is necessary to provide training in water supply techniques and procedures. Many pump operators simply cannot draft in even the simplest water supply situations. Very few training opportunities exist for water supply officers to hone their water shuttle or practical operations skills. This type of training should be in every fire department's long-range plan for improvement.

Water supply for large fires is rarely accomplished solely through the efforts of one fire department. Because no one fire department has to invest the money to buy all of the apparatus and equipment that will be needed, mutual aid is the most cost-effective way, and, indeed, is essential to providing an adequate level of fire protection throughout an area. ISO recognizes the value of mutual aid in attaining the most fire protection at the least cost and its role in handling extraordinary situations. While mutual aid resources can be given credit in an ISO evaluation, an automatic aid factor is applied to the apparatus and personnel regularly responding from neighboring departments. The maximum credit that can be given for an automatic aid department is 90% of the points it would score in an evaluation. To get the maximum credit, a number of requirements must be met, including method of dispatching, and radio frequencies. Automatic aid companies have to be dispatched automatically and simultaneously with the first-due company. Also a minimum of four half-day drills with automatic aid companies must be conducted and recorded each year. These drills offer a good opportunity to test some of the preplans that have been developed, and are essential if automatic aid companies are to work as a team on the fireground. They should be planned well in advance, participation should be recorded, and the results should be documented in the training records of both companies. These training sessions also can be designed to meet the requirements of the Superfund Amendments and Reauthorization Act (SARA) of 1986.

Training is the key to improving the level of fire protection in a community, and also is the best way to ensure that a fire department's customers get the most fire protection for their money.

LONG-RANGE BUDGETING

One important aspect of long-range planning is determining budgetary requirements. By including a cost estimate for each recommendation, the fire department and its governing body can prioritize needs. Two long-range budgets should be prepared: an operating budget and a capital budget. The operating budget includes normal operating costs, i.e., salaries, buildings, utilities, insurance, repairs, fuel, and other operating expenses, while a long-range capital budget consists of funds intended for major expenditures, e.g., building construction, building modifications or expansion, apparatus, and equipment. Since the normal life of a pumper or ladder truck ranges from 20 to 30 years, a long-range budget should project at least 20 years into the future. The long-range budget will need to be adjusted to change with responsibilities and needs, but a 20-year view allows the fire department to plan for funding improvements instead of reacting with crisis management.

FUNDING

After needs are determined, a means of obtaining the required funds should be identified. Fire department funding can come from tax revenues, fundraising, contributions or long-term debt financing.

TAXES

Paid fire departments generally are funded entirely by tax revenues. There are two ways local governments can budget for fire protection. The fire department can be a line item in the budget with all expenses paid from general fund revenues. This method of budgeting gives the local government the advantage of allocating funds on an as-needed basis. The disadvantage is that the fire department must compete with other government agencies at budget time each year; this makes long-range planning difficult.

Some local governments have imposed a dedicated fire tax or fee to provide funding for fire protection. The fire tax can take the form of a general purpose levy across an entire city or county, or special-purpose tax districts can be formed. The problem with forming a number of special-purpose tax districts within a county or other subdivision is that the level of fire protection varies widely. Districts which have a number of high value properties can raise all the money needed with a relatively low tax rate; however, in sparsely settled areas even a very high tax rate yields insufficient funds. By imposing a uniform tax rate across a wider base, money can be allocated to each fire department according to its needs, without imposing an undue burden on any of the taxpayers.

Some form of property tax is the most equitable way to fund fire protection. While the fire department makes every effort to save lives, the fact is that most fire deaths occur before the first fire engine turns a wheel. The primary function of the fire department is to save property. The best measure of the benefit a taxpayer receives from the fire department is the value of the property being protected. With a property tax, the owners of property with the highest value pay the most toward the operation of the fire department. From the fire department's standpoint, the dedicated fire tax has two major advantages:

- First, the dedicated fire tax generates a predictable income and provides a base for long range planning.
- Second, it has an inflation growth factor built in to the tax base. As the value of the properties being protected increases, the funding for fire protection automatically follows.

CONTRIBUTIONS AND FUNDRAISING

Volunteer fire departments traditionally have depended primarily on fundraising activities and contributions from their constituents to provide operating funds. Many of them have been unable to raise enough money to operate safely and effectively and have been looking for greater support from the local government for two reasons:

- First, with new technology and increasing government regulation, the costs of operating a fire department have increased dramatically.
- Second, firefighter certification, hazmat requirements, emergency medical system needs, and other training standards volunteers are required to meet use an ever increasing amount of their time, leaving less for fundraising and other activities.

Volunteers donate their time and risk their lives serving their community. The least the taxpayers and residents of the area can do is to provide them with the tools they need to do their job. The primary responsibility of the fire department is to train, prepare for and provide emergency services. When volunteers are distracted by the need to engage in fundraising activities and to solicit contributions, the quality of protection they provide suffers. In the long run, it is the community that is short-changed by a failure to provide the needed funding.

COMPREHENSIVE PLANS

The objectives of all fire departments are to control emergency situations, to prevent loss of life, and to minimize property damage. The goal of fire suppression is to apply enough water to a fire to bring it under control quickly. All of the concerns expressed in this chapter contribute to a department's ability to do this.

An in-depth evaluation should be done at least once every five years, and a comprehensive plan for fire protection created. Once a comprehensive plan has been developed, a plan for implementation should be adopted. The plan for implementation should include immediate objectives, long-range goals, and a budget to fund them. It also should include benchmarks to be met, with regular reviews to determine that progress is being made.

3. Risk Analysis In Fire Protection Planning

COMPREHENSIVE PLANNING for fire protection begins with determining the water supply requirements of the area to be protected. One way to determine these needs is to divide the fire department's area of responsibility into Fire Management Areas (FMAs). Defined by the National Fire Academy in its Community Fire Defense: Challenges and Solutions curriculum, FMAs are created by grouping occupancies that have similar fire protection needs within natural or manmade boundaries. Alarm assignments then can be made that will ensure an adequate response to the types of problems that can be expected within each FMA. An estimate of Needed Fire Flow (NFF) is made for each FMA and the total amount of water required, as well as the maximum delivery rates, are calculated.

Once the boundaries of the FMA have been established, it's necessary to determine the water supply needs of the type of structures to be protected. During its evaluation, the Insurance Services Office (ISO) bases the needed fire flow on the fifth largest building in the area under consideration. This approach provides enough water to handle most situations, sets a goal for the fire department that is reasonable and attainable, and minimizes needless response of manpower and equipment.

Water supply planning for the largest buildings and target hazards is handled on a case-by-case basis with individualized plans for each. Many fire departments establish special alarm assignments with enough apparatus on the initial response to provide the needed water supply. Otherwise there would be a delay between the first alarm unit's arrival and the dispatch of additional water supply units. If the first engine companies carry adequate water they can make an aggressive initial attack.

RISK ANALYSIS

Among the methods of calculating needed fire flow are those of ISO, NFPA, and the National Fire Academy. For planning purposes, the one that seems the simplest to use is the method suggested in NFPA 1231, *Standard on Water Supplies for Suburban and Rural Fire Fighting*.

The NFPA formula to determine the total water supply requirement is

$$\frac{V}{OHC} \times CCN \times EF = TWS \text{ (Total Water Supply)}$$

where

V = Total volume of the building in cubic feet
OHC = Occupancy Hazard Classification
CCN = Construction Classification Number
EF = Exposure Factor
TWS = Total Water Supply needed for the structure

TOTAL WATER SUPPLY

Calculated using the NFPA method, the Total Water Supply (TWS) provides an estimate of how much water will be required to fight a fire in the building being surveyed if the building is fully involved. In general, the larger the structure, the more water needed. A minimum water supply of 2,000 gallons, where there are no exposures, or 3,000 gallons, where unattached structures are located within 50 feet of the building being evaluated, is required.

VOLUME OF THE STRUCTURE

The TWS calculation is based on full involvement of the structure, and uses the total volume of the building. This calculation should include the basement, attic, and any attached structures the fire could affect. Since this figure is only an estimation, precise measurements are not required. You can round dimensions off, and use the average height of the building when there are peaked or sloping roofs. When multi-story buildings are being evaluated, all floors should be included.

OCCUPANCY HAZARD CLASSIFICATION

The total water supply needed is based on a preplanned estimate of potential fire volume. In addition to the size of the building, its occupancy type and method of construction contribute to the extent of the fire hazard. A large building with a light hazard use will present fewer problems, and require a lower TWS, than a much smaller building that contains hazardous materials or large quantities of flammable or combustible materials.

Occupancy Hazard Classification (OHC) numbers range from three to seven. Lower numbers indicate a greater degree of hazard. The OHC numbers, along with some examples of each, are:

OHC 3—SEVERE HAZARD OCCUPANCIES
   Explosives manufacturing and storage
   Lumberyards
   Baled straw or hay in bales
   Plastics manufacturing and storage
OHC 4—HIGH HAZARD OCCUPANCIES
   Building materials storage
   Department stores
   Repair garages
   Warehouses
OHC 5—MODERATE HAZARD OCCUPANCIES
   Farm storage buildings
   Libraries
   Restaurants
   Unoccupied buildings
OHC 6—LOW HAZARD OCCUPANCIES
   Churches
   Gasoline service stations
   Telephone exchanges
   Funeral homes
OHC 7—LIGHT HAZARD OCCUPANCIES
   Apartments
   Hotels
   Schools

CONSTRUCTION CLASSIFICATION NUMBER

The total water supply requirement for a specific building is influenced strongly by the type of construction. A Construction Classification Number (CCN) between 0.5 and 1.5 is assigned to the structure being evaluated. The CCN is based on the materials used and the method of construction. The higher the CCN, the greater the fire risk. Simplified descriptions of each classification are:

CCN 0.5—FIRE-RESISTIVE CONSTRUCTION
Noncombustible materials; all metal members protected from heat; major structural members designed to withstand collapse.

CCN .75—NONCOMBUSTIBLE CONSTRUCTION
All structural members (including floors, walls, and roofs) noncombustible materials.

CCN 1.0—ORDINARY CONSTRUCTION
Exterior walls of noncombustible material; structural members wood or other combustible material.

CCN 1.5—WOOD FRAME CONSTRUCTION
Exterior walls and structural members wood or other combustible material.

Note: Dwellings are assigned a CCN no higher than 1.0, even if they are of wood frame construction.

EXPOSURE FACTOR

When any portion of an unattached structure is located less than 50 feet from the building being evaluated, is larger than 100 square feet, and will be exposed to heat from the fully involved primary risk, the potential exists for the fire to spread and involve both buildings. If two buildings are involved, there will be a corresponding increase in the volume of fire that can be expected and the amount of heat generated. The TWS must be increased to allow for the exposure hazard. Buildings smaller than 100 square feet are considered an exposure problem if they house dangerous materials or are another type of high-hazard occupancy (explosives storage); the exposure factor should be applied to the calculations in these cases. After you calculate the TWS requirement for the primary risk, based on the volume of the structure, the OHC factor, and the CCN factor, multiply the final result by 1.5, the exposure factor (EF) to determine the total water supply needed for the structure, including the exposure.

CALCULATING THE TOTAL WATER SUPPLY

Let's look at an example of calculating total water supply requirements. You have a barn or other storage building that is 80' × 60', with a height of 25' to ridgepole and 15' to the eaves, of wood frame construction. There is a 20' × 40' equipment shed located 30 feet from the storage building. Begin your calculations by determining the total volume of the building in cubic feet.

L (length) × W (width) × H (Average Height) = V (Volume cubic feet)

To determine total volume, the average height of a building with a peaked roof is calculated by:

$$\frac{Hp \text{ (peak)} - He \text{ (eaves)}}{2} + He \text{ (eaves)} = H(\text{Average Height})$$

The two equations are combined to give the total volume of the building:

$$80 \text{ (L)} \times 60 \text{ (W)} \times \frac{(25 - 15)}{2} + 15 = 96{,}000 \text{ Cubic Feet}$$

Take the building's total volume (96,000 cu. ft.) and factor in:
 Moderate Hazard Occupancy—OHCN 5
 Wood Frame Construction—CCN 1.5
 Exposure factor—1.5
 to calculate the Total Water Supply.

$$\frac{96000 \text{ (V)}}{5 \text{ (OHC)}} \times 1.5 \text{ (CNN)} \times 1.5 \text{ (EF)} = 43{,}200 \text{ (TWS)}$$

DELIVERY RATE

Even if the total water supply requirement can be met for a given FMA, the location of the water supply must be considered. Unless the storage is located on site, or is connected to an approved fire hydrant system, it will be necessary for the fire department to transport the water from the supply location to the point of need. In NFPA 1231, guidelines for minimum fire flow requirements, based on required total water supply are specified.

Table 3.1. Delivery Rate Based on Required Total Water Supply

Total Required Water Supply	Required Delivery Rate
Less than 2,499 Gallons	250 GPM
2,500 to 9,999 Gallons	500 GPM
10,000 to 19,999 Gallons	750 GPM
20,000 Gallons or more	1,000 GPM

The guidelines in Table 3.1 do not go far enough! Experience has shown that many structures and situations have the potential to exceed a 1,000-GPM flow. In 1231, the NFPA points out that large or special fire protection problems could require greater flows, yet seems to imply that suburban and rural fire departments would not be able to deliver them. However, many fire departments have disproved this by providing fire flows greater than 1,000 GPM over significant distances. These fire departments credit advance planning and efficient use of all available resources for these accomplishments. If an exposure presents special problems (e.g., flammable liquid or gas storage, or explosives), you can estimate additional fire flow by considering the number and type of attack lines that would be required to protect the exposures, and adding the extra GPM to the delivery rate previously determined.

APPLICATION RATE

Total water supply requirements as well as delivery rates are based on the maximum amount of water that would be required to control a fully involved structure fire. Successful extinguishment depends on applying the water to the fire rapidly enough to absorb heat faster than it is being generated. If the suppression efforts of initial engine companies are successful, the fire will be controlled within minutes of their arrival, and can be prevented from extending to the portion of the structure that was not originally involved.

One commonly used formula for determining application rates for rapid fire control is

$$\frac{\text{length} \times \text{width} \times \text{height}}{100} = \text{GPM}$$

We can apply this formula to our storage building example:

$$\frac{80 \times 60 \times 20}{100} = 960 \text{ GPM}$$

The needed fire flow of 960 GPM could be met with the recommended delivery rate of 1,000 GPM.

If we double the size of the building to 160' × 60', but leave all other provisions unchanged, the total water supply requirement would rise to 86,400 gallons. Calculate the fire flow for the larger structure.

$$\frac{160 \times 60 \times 20}{100} = 1,920 \text{ GPM}$$

The maximum delivery rate of 1,000 GPM recommended in Table 3.1 would supply only approximately half of the amount of water that would be needed to control a fire in this building. This example illustrates graphically the need for specific plans to deliver water above and beyond the maximum flow specified in NFPA 1231. This is especially true when you must supply the number of attack lines that would be necessary for fire control in large structures with special fire protection problems.

INITIAL ATTACK

We have learned that total water supply, delivery rate, and application rate calculations are based on the total volume of the structure. However, many structures are divided into areas by fire walls or other types of compartmentation that provide some degree of isolation from the rest of the building. It is in one of these areas that the fire department can reasonably expect its first alarm response to deal with a fire. Based on this expectation, the maximum application rate needed for an effective initial attack would be determined by the dimensions of the largest open space, rather than on those of the complete structure, for prefire planning purposes.

Take, for example, a one-story school building that measures 200' × 120'. It has an average height of 15 feet, but includes an auditorium that is 50' × 50' with a ceiling height of 20 feet. The building also has a number of individual classrooms. The building is of ordinary construction; there is no exposure within 50 feet. Total water supply needs for the entire building, based on the formula used in NFPA 1231, would be

$$\frac{200 \times 120 \times 15}{7 \text{ (OHCN)}} \times 1(\text{CCN}) = 51{,}429 \text{ Gallons}$$

This is more than twice the TWS that would require a 1000-GPM delivery rate as specified in Table 3.1.

The needed fire flow, based on the application rate formula, would be

$$\frac{200 \times 120 \times 15}{100} = 3{,}600 \text{ GPM}$$

If the fire is confined to the largest undivided space—the auditorium—the flow rate for initial attack, based on the application rate formula, would be reduced

$$\frac{50 \times 50 \times 20}{100} = 500 \text{ GPM}$$

This example shows that the delivery rate specified in Table 3.1 would be grossly inadequate if the fire were allowed to progress to total building involvement before the arrival of the first alarm units. On the other hand, in an occupancy of this type, it is probable that the first alarm units would encounter a fire that is still confined to one of the rooms. Even if the largest room, the auditorium, were fully involved, a delivery rate of 1,000 GPM should easily supply the initial attack lines.

PREFIRE PLANNING

Prefire plans for target hazards and special risks should consider the expected fire flow requirements for an effective initial attack without ignoring the possibility of total involvement. First alarm assignments should be able to supply enough water to handle the initial attack, with subsequent alarms set up to deliver enough water to meet the total water supply requirements for the maximum possible volume.

With this in mind, fire departments need detailed plans for all target hazards and special risks within their area of responsibility. Not only does a plan meet ISO evaluation requirements, but it allows the department to establish alarm assignments that are targeted for special risks.

Each large structure or special risk should be evaluated on a case-

Figure 3-1. Old residence converted to a sheet metal shop in Owings, MD. with flammable liquid storage in background.

by-case basis. Figure 3-2 is an example of a form that can be used to gather data for planning. In this example, the plan is for a small commercial building located in a rural area beyond the reach of any water system. The size of the building, its construction classification and occupancy hazard, and any information about exposure hazards, are used to calculate the maximum total water supply that may be needed. The commercial building used in this example has a total volume of 40,800 cubic feet, is of wood frame construction, and has a moderate occupancy hazard. The exposure hazard is another building less than 50 feet from the structure. The water supply calculation indicates that a total supply of approximately 18,360 gallons would be needed. Applying the NFPA guidelines for needed fire flow, a fire in this structure could require as much as 750 GPM.

The degree of life hazard is significant to the fire department. In this case, an apartment on the second floor could present a potentially difficult rescue situation, with resulting loss of life, if a fire were to progress rapidly.

Initial response to a target hazard generally involves more than one engine company. The plan should note access problems, and specify any special arrangements the initial engine company would have to make to ensure access to the building by later-arriving ladder trucks, attack units, and water supply equipment.

FIRE RISK ANALYSIS WORKSHEET

FIRE DEPARTMENT __North Beach Vol. Fire Dept.__ DATE __May 26, 1989__

DESCRIPTION AND LOCATION: Krick Sheet Metal Company
                            Rt 778 and Thomas Ave.
BOX 106 - OWINGS, MD.   Owings, Md.

FIRE FLOW CALCULATIONS:

    A.   Occupancy Hazard Classification   __Moderate OCHN 5__

    B.   Construction Classification   __Wood Frame CCN 1.5__

    C.   Dimensions, Largest Building   Length __34__

                                                    Width __40__

                                                    Height __30__

    D.   Volume of Primary Risk   __40,800__

    E.   Exposures within the danger zone:

Chesapeake Animal Clinic, less than 50 feet away and the Southern Maryland Oil Bulk Plant with above ground flammable liquid storage within 100 feet of the primary risk building.

$$\frac{40{,}800\ (C)}{5\ (A)} \times 1.5\ (B) \times 1.5\ (D) = 18{,}360\ \text{Gallons (TWS)}$$

TWS __18,360__   Needed Fire Flow __750 GPM__

Figure 3-2. Risk Analysis Worksheet for Krick Sheet Metal Company in Owings, MD.

WATER SUPPLY AVAILABILITY

Once you have determined the needed fire flow for a specific risk, your preplan should specify its source. The water supply availability worksheet associated with the structure described in Figure 3-2 is shown in Figure 3-3. This building is located in a small community with no public water system, where the fire department has to depend on tankers to haul the water. There are two ponds located within two miles of this target hazard that can be used for water supply. Since a fire flow greater than 500 GPM is needed and six tankers would be used in a water shuttle to provide it, the preplan calls for setting up tanker fill points at both locations. WS 106-1 is a large storm water management pond located on an improved gravel road approximately one-half mile from the fire location, with an estimated roundtrip travel time of three minutes. First alarm tankers would load at this water supply point. A second water supply point located one and a half miles away, over a paved secondary road, has an estimated travel time of seven minutes. A mutual aid engine company would be dispatched to set up a draft at this location, and the second alarm tankers would be assigned to load there.

MOBILE WATER SUPPLY APPARATUS RESPONSE

In the example shown as Figure 3-3, the initial engine (E-12) with 1,000 gallons of water would be dispatched, plus three tankers. The first tanker (Tkr 5) ideally would arrive within two minutes of E-12 (+2 on the chart). The remaining tankers (Tkr 2 and AA1) would arrive approximately 11 minutes later (+13 on the chart). The second alarm assignment includes another three tankers, but with a longer travel time, they would arrive as much as 38 minutes after the arrival of the engine company.

The usable water, the handling time, and the expected travel time all were used to calculate a GPM rating for each tanker loading at each of the water supply points that will be used for this plan. Chapter 5, Testing Water Supply Apparatus, explains how to determine each tanker's continuous water supply.

Since the three-minute estimated travel time to the storm water management pond (WS 106-1) is shorter than the seven-minute trip to SSS WS502-1, each tanker could deliver nearly 50% more water from the closest water supply point. Based on this chart, the first alarm tankers would be able to supply 771 GPM, which is more than the base fire flow figure arrived at in the calculations. The second alarm tankers, using the more distant water supply point, would then deliver another 474 GPM, more than enough to supply the initial attack lines and provide for the exposure problem.

FIRE FLOW WORKSHEET

After you have determined the needed fire flow, and evaluated whether the fire department can supply it, a fire flow worksheet can be

RISK ANALYSIS - Krick Sheet Metal Company, Owings, MD.

REQUIRED FIRE FLOW - 750 GPM

		AVAILABLE WATER SUPPLY POINTS			
	TRAVEL TIME (MIN)	LOCATION	CAPACITY	SUPPLY	
A	3.0	WS 106-1 off RT 260	Unlimited	Pond	
B	7.0	SSS 502-1, Mt Harmony Rd across from school	Unlimited	Pond	
		TANKER RESPONSE			
ETA	HANDLING TIME	TANKER	CAPACITY	GPM A	GPM B
1		Engine 12	1000		
2	6	Tanker 5	1988	221	153
13	6	Tanker 2	3217	357	247
13	11	Tanker AA1	2700	193	150
		FIRST ALARM TOTAL		771	550
19	6	Tanker PG 23	2300	256	177
30	4	Tanker 81	1463	209	133
39	4	Tanker 22	1800	257	164
		SECOND ALARM TOTAL		722	474
		TOTAL AVAILABLE FLOW		1,493	1,024

NOTES:
1. Life hazard in apartment on second floor over metal shop.
2. Southern Maryland Oil has 160,000 gallons of fuel oil and 40,000 gallons of K-2 fuel in above ground storage.
3. Parking area to side of building could be used for a dump site to operate a water shuttle.

Figure 3-3. Water Supply Availability Worksheet for Krick Sheet Metal Company in Owings, MD.

prepared. This worksheet lists the fire flow that can be sustained, and more importantly, the timeframe in which it can be initiated. Figure 3-4 is an example of a computer-generated fire flow worksheet that was developed from the risk analysis information in this example.

As shown on the worksheet's "arrive" column, the initial engine company brings 1,000 gallons of water. One minute later, the first tanker (Tkr 5) arrives with 1,811 gallons of usable water (Usable Water, Initial). Tkr 5 dumps its load into a portable tank, and leaves three minutes later to get another load of water. After refilling at WS 106-1, Tkr

	Krick Sheet Metal Co. Owings, MD. WS 106-1, SSS 501-1					
	BOX 106, OWINGS TOWN CENTER					
	TANKER MOVEMENT		Tanker	Avail	Fire	Net
MINUTE	Arrive	Depart	Arrivals	Accum.	Flow	On Hand
1	E12		1000	1,000		1,000
2	Tkr5		1811	2,811	200	2,611
3				2,811	200	2,411
4				2,811	200	2,211
5		Tkr5		2,811	200	2,011
6				2,811	200	1,811
7				2,811	200	1,611
8				2,811	200	1,411
9				2,811	400	1,011
10				2,811	400	611
11				2,811	400	211
12	Tkr5		1988	4,799	400	1,799
13	Tkr2,AA1		5489	10,288	400	6,888
14				10,288	400	6,488
15		Tkr5		10,288	1,000	5,488
16		Tkr2		10,288	1,000	4,488
17				10,288	1,000	3,488
18				10,288	1,000	2,488
19	PG23		2300	12,588	1,000	3,788
20		AA1		12,588	1,000	2,788
21	Tkr5		1988	14,576	1,000	3,776
22		PG23		14,576	1,000	2,776
23				14,576	1,000	1,776
24	Tkr2	Tkr5	3217	17,793	1,000	3,993
25				17,793	1,000	2,993
26				17,793	1,000	1,993
27		Tkr2		17,793	1,000	993
28	AA1		2700	20,493	1,000	2,693
29	PG23		2300	22,793	1,000	3,993
30	Tk81		1444	24,237	1,000	4,437
31	Tkr5		1988	26,225	1,000	5,425
32		PG23,TK81		26,225	1,000	4,425
33				26,225	1,000	3,425
34	Tkr2	Tkr5	3217	29,442	1,000	5,642
35		AA1,PG23		29,442	1,000	4,642
36				29,442	1,000	3,642
37		Tkr2		29,442	1,000	2,642
38				29,442	1,000	1,642
39	Tkr22		1800	31,242	1,000	2,442
40	Tkr5		1988	33,230	1,000	3,430
41	Tk81		1463	34,693	1,000	3,893
42		Tkr22		34,693	1,000	2,893
43		Tk81,Tkr5		34,693	1,000	1,893
44	AA1		2700	37,393	1,000	3,593
45	PG23		2300	39,693	1,000	4,893
46				39,693	1,000	3,893
47	Tkr2		3217	42,910	1,000	6,110
48		PG23		42,910	1,100	5,010

Figure 3-4. Fire Flow Worksheet for Krick Sheet Metal Company in Owings, MD.

5 returns to the dump site seven minutes later with 1,988 gallons of water (Usable Water, Shuttle). Now, according to Column 4 (Avail. Accum.), a total of 4,799 gallons of water has been delivered to the fire scene. One minute later, two additional tankers (Tkr 2 and AA1) arrive, bringing another 5,489 gallons (Usable Water, Initial), for which a second portable tank will be set up. The tankers dump their load, and join the water shuttle.

For this plan, the 1988 edition of NFPA 1410, *Standard on Initial Fire Attack*, was used as a guideline. The plan calls for putting two attack lines (flowing 200 GPM total) in service within one minute of arrival. Two minutes later, a backup line is added (flowing 200 GPM) for a total flow of 400 GPM. This 400 gallons, listed in the "fire flow" column, is subtracted from the water that was brought to the scene, as shown in the "avail. accum." column, and the "net on hand" is listed in the last column. Given the travel time for the second- and third-due tankers, the worksheet indicates that 15 minutes would elapse from the time of arrival of the initial engine company before a master stream could be put into service and sustained without interruption. Even with variances in the amount of water on hand at the fire scene, a minimum flow of 1,000 GPM could be sustained for at least the first 48 minutes of the attack. The 15-minute delay in reaching the full 1,000-GPM flow falls within the permissible timeframe used by the ISO when it assigns a community's PPC rating using (Fire Department Water Supply) techniques.

IMPLEMENTATION

A detailed plan should be part of the information book that is carried in each piece of apparatus and in the chief's car. It should be readily available to any incident commander and be kept at the incident command post. Each officer, from the individual company officer to the department's chief, should know the plan's format, ideally from having participated in the planning process.

The preceding risk analysis methodology showed that assistance from mutual aid companies is required to control emergencies that involve target hazards and special risks. Any fire departments asked to provide mutual aid should be given the same information that is available to the initial engine and ladder companies. Mutual aid drills are essential if operations of this magnitude and complexity are to be successful. Joint training sessions can center on plans for target hazards, followed by a "dry run" with all of the units that will be dispatched on an alarm participating. The "dry run" highlights any flaws in the plan, enabling the department to seek alternatives and make adjustments before actual emergencies occur.

Prefire planning target hazards should be an ongoing process. To help in comprehensive planning, a plan for one particular building can serve as a model for similar structures in the same general vicinity. Over a period of time, all major hazards should be evaluated and a definite preplan prepared for each one.

Remember, a preplan is only as good as its implementation.

4. The Use of Fire Hose in a Water Delivery System

AMONG THE RECENT ADVANCES in fire hose manufacturing is the switch to synthetic fabrics instead of cotton for the outer jacket. While synthetic fabrics are an improvement because of better wear and flexibility, they also make it difficult to determine friction loss. Synthetic fabrics vary widely in both elasticity and strength, creating different friction loss characteristics for the same type and size of fire hose when it is made by different manufacturers. This is especially significant in large diameter hose (LDH). For this reason, friction loss calculations can only approximate the amount of loss that can be expected. Actual flow and pressure tests have to be made if an exact friction loss figure is needed for a particular hose layout.

Through the 1940s, the standard supply line for the fire service was rubber-lined fire hose with a double cotton jacket. When some synthetic materials became readily available in the 1950s, manufacturers introduced a variety of different types of hose to the fire service market. Although today's fire departments have a vast new array of products available to them, many departments are not able to use them effectively. Some are reluctant to try a new product, while others see the new hoses as a panacea for all types of fires and all types of incidents. This faulty logic occurs when the characteristics of a given hose type and its advantages and disadvantages when used in a water delivery system have not been considered.

Fire hose serves two separate and distinct functions in a water delivery system. **Attack** lines are used to move water from the attack pumper to the nozzle for application to the fire. **Supply** lines are used to bring the water from the source to the attack pumper and must provide all of the water needed to control the fire. Each of these functions needs to be considered separately.

ATTACK LINES

Hose reels that are equipped with ¾″- and 1″-rubber hose should never be used as fire attack lines and will not be considered for this use here. The maximum flow these "booster lines" can yield is insufficient to handle any significant volume of fire. Many departments have made the mistake of attacking an incipient fire with booster lines, only to be outrun by the fire. In some cases, buildings have been lost before larger lines could be put into service.

For some years, preconnected 1½″- and 2½″-lines have been the standard for initial attack by most fire departments. Many departments now use either 1¾″- or 2″-hose as a replacement for one or both of these. In deciding whether to make the change, some of the characteristics of various hoselines need to be considered.

The most obvious reason to change hose size is the difference in weight. Attack lines need to be mobile, and the weight of a charged hoseline can limit its use. The higher the operating pressure at the pump panel because of the friction loss, the more rigid and difficult it is to advance the attack line. This is especially troublesome when working in the confined spaces typical in interior attacks. Often overlooked is the maximum GPM that can be supplied through a given hose size within the limits of a safe operating pressure.

Tables 4.1 and 4.2 summarize some of the characteristics you should consider when selecting attack line fire hose.

Table 4.1. Weight of a 100-Foot Section of Fire Hose Full of Water

Size of the Hose	Capacity in Gallons	Approximate Weight		
		Water	Hose	Total Weight
3″	36.8	307	164	471
2½″	25.5	212	130	342
2″	16.4	136	100	236
1¾″	14.3	120	85	205
1½″	9.2	77	70	144

NOTE: The weight of the hose will vary with the material and method of construction; these figures are typical of some of the products most commonly used in the fire service.

The Use of Fire Hose in a Water Delivery System

Table 4.2. Friction Loss in a 100-Foot Section of Hose According to Flow

Size	250 GPM	200 GPM	150 GPM	120 GPM	90 GPM	60 GPM
2½"	13.4	8.9	5.3	2.6	1.5	.7
2"	40.8	27.2	16.2	7.8	4.6	2.3
1¾"	79.6	53.0	31.5	15.3	9.0	4.3
1½"	174.2	114.5	68.2	33.0	19.5	9.2

The amount of change in friction loss as flow increases or decreases in smaller, more mobile handlines is a problem. Friction loss changes according to the amount of water flowing. If a manually adjustable, variable flow nozzle is used with the fire pump applying constant pressure to the attack line, excessive changes in nozzle pressure will occur when the nozzle setting is adjusted. When the flow in GPM goes up, friction loss in the hose increases, and the nozzle pressure is reduced. Table 4.3 shows the amount of change that can be expected as the rate of flow is varied across some typical settings of variable flow nozzles. The calculations in this table are based on the use of a 250-foot attack line. For longer hoselines, the variations in pressure that occur with changes in the GPM setting will be greater than those shown in the table. Shorter lines will reduce the amount of change to be expected.

Table 4.3. Nozzle Pressure Variations with Flow Changes in a 250'-Hoseline

Hose Size	Pump Pressure	Nozzle Pressure with Changes in GPM						Chan in Pres
		250 GPM	200 GPM	150 GPM	120 GPM	90 GPM	60 GPM	
2½"	122	89	100	109				30
2"	168	66	100	127				61
2"	112*				92	100	106	14
1¾"	123*				84	100	112	28
1½"	149*				65	100	125	60

NOTE: Pressures marked with an * assume the use of a 1½"-adjustable nozzle with the maximum flow setting of 120 GPM.

While the use of an automatic nozzle will eliminate these wide variations in nozzle pressure, it may require the use of excessively high pressures at the pump panel to attain the desired flow rate. Additives can reduce friction loss and make the use of smaller attack lines more practical; however, the chemicals are expensive, and most fire departments are reluctant to use them in pump operator training. When personnel have to operate a pump under actual fireground conditions, they will be faced with an unfamiliar situation and will not be able to maintain the desired nozzle pressure on the attack lines. While there are certain advantages to using additives to minimize friction loss, their use is not cost effective for most fire departments.

Many fire departments are not aware of the limitations that are imposed on automatic nozzles supplied by 1¾″- or 2″-hose. Based on information provided by the manufacturer, Table 4.4 shows the maximum flows that can be expected when 150′- or 200′-attack lines are used with one of the most popular automatic nozzles. These figures would probably be typical of the limitations imposed by friction loss on the flexibility of this type of device.

Table 4.4. Required Pressure at the Pump for Different Flow Rates Using an Automatic Nozzle

Flow Rate in GPM	Required Engine Pressure							
	1½″-Hose		1¾″-Hose		2″-Hose		2½″-Hose	
	150′	200′	150′	200′	150′	200′	150′	200′
150	200	235	140	150	125	135	110	115
200	*	*	170	190	145	160	115	120
250	*	*	200	235	170	190	125	130
300	*	*	250	*	200	230	130	140
350	*	*	*	*	230	*	140	155

NOTE: Pressures marked with * would exceed 250 PSI, the maximum operating pressure for a Standard fire pump.

Table 4.4 makes it clear that the maximum flow attainable from an automatic nozzle decreases rapidly as the attack line gets longer. Some fire departments carry preconnected lines as long as 400 feet to use in large commercial or industrial building fires. At the same time, there is a tendency to use 2″-lines as a substitute for 2½″-hose to increase the mobility of the attack lines. The increased friction loss in the longer, smaller diameter hose reduces the maximum flow, making it inadequate to handle the volume of fire likely to be encountered. The only way to take advantage of the versatility of the automatic nozzle and realize its full potential with long attack lines is to use larger hose.

FIREGROUND HYDRAULICS

A good standard operating practice is to limit the maximum friction loss in attack lines to 30 PSI per 100-foot section. With an optimum nozzle pressure of 100 PSI, the pump operator could supply a 300-foot line with a pump pressure of 190 PSI. At 190 PSI, the pump operator would be able to increase the pressure enough to allow for head pressure on the upper floors of a building without exceeding a safe working pressure. Typical preconnected lines of 150 or 200 feet could then be used with an operating pressure of 160 PSI, well below the safe operating pressure of the fire hose (as tested) and low enough to provide maximum mobility of the attack lines. The use of solid bore nozzles with a recommended operating pressure of 50 PSI or low-pressure fog nozzles with an operating pressure of 75 PSI would reduce the required pump pressure, increase the maximum flow, or permit the use of longer attack lines.

Using this rule of thumb, the maximum flows from various attack lines would be:

Table 4.5. Maximum Flow From Attack Lines with 30 psi Friction Loss per 100-Foot Section

Size of the Hose	Maximum Flow
1½″-Line	120 GPM
1¾″-Line	150 GPM
2″-Line	200 GPM
2½″-Line	375 GPM
3″-Line	550 GPM

NOTES:
- While 3″-lines generally are not used as attack lines and are too heavy to provide much mobility as handlines, they often are used to supply master stream devices. The maximum flow rates for 2½″- and 3″-lines in this table assume the use of multiple lines to a master stream device.
- When 2½″-hose is used for a handline with a manually adjusted nozzle, the maximum flow generally will not exceed 250 GPM because most 2½″-nozzles have a maximum flow rate of 250 GPM. Another limitation is the reaction that occurs when water leaves the nozzle, i.e., automatic nozzles may have a maximum flow rate of 350 GPM, but it would be difficult, if not impossible, to maintain control of a 2½″-handline discharging more than 250 GPM.
- The proposed maximum flows for 2½″- and 3″-**attack** lines differ significantly from those expected for **supply** lines. A typical supply line is much longer than an attack line, and the increased friction loss in the additional hose sections limits the maximum flow rate.

SUMMARY OF ATTACK LINE CONSIDERATIONS

Since the mobility of attack lines is affected by operating pressure and hoseline weight, the length of the preconnected lines, the expected flow rates, and the type of nozzle are all significant factors in determining the best size to use for a particular situation. There is no "all-purpose" hose. Attack lines have to match the needs of the area served and the type of structures to be protected, and decisions about them should not be based on reluctance to change, the constant search for something new, or be affected by prejudice. Newer is not always better, nor is repeating past practices necessarily the best approach.

SUPPLY LINES

Supply hose is defined as hose used to transport water from the water source to an attack pumper, and it is measured by a completely different standard than attack hose connected to the discharge side of the pump. Attack lines are designed to supply a limited amount of water to some type of water flow device over relatively short distances at the desired operating pressure. Supply lines generally are required to supply larger amounts of water to the pump over longer distances, with only a minimum residual pressure required at the intake of the attack pumper. At times, one supply line is called upon to provide enough water to operate several attack lines connected to the same pumper. What primarily limits a particular hose layout's ability to do this is the friction loss that occurs as the water moves through the hose from the supply point to the fireground. One of the most common problems engine companies have when attempting to make an effective attack is that their supply lines are not large enough to supply the water as fast as it is needed to control the fire.

Most supply line calculations are based on assuming a maximum distance between pumpers of 1,000 feet. Following are several reasons why 1,000 feet is used as a standard.

According to hydrant spacing requirements established by the Insurance Services Office (ISO), any protected risk should be within 1,000 feet of a hydrant.

All pumpers are required to carry at least 1,000 feet of supply line to meet the ISO's Rating Guide standards.

Pumpers are rated at a maximum net pump pressure of 150 PSI when they are supplying 100% of their rated capacity.

To operate within these parameters, fire hose used for supply line should be sized to limit total friction loss to 15 PSI per 100 feet based on the maximum flow rate of the flow devices to be used. If friction loss in each 100-foot section of hose is 15 PSI or less, then the required net pump pressure to overcome the losses when supplying a 1,000-foot hoseline will not exceed 150 PSI. Any pumper in good condition should be able to provide this. When there is a significant difference in elevation between the two ends of the supply line, the maximum length must be reduced to allow for head pressure. The head pressure is created by the weight of the water in the line and must be compensated for by the pumper that is supplying it.

2½-INCH HOSE

Traditionally, 2½"-hose has been the standard for supply line in the fire service. The ISO's 1980 Fire Protection Rating Schedule calls for 1,200 feet of 2½"- or larger hose on each pumper. If we apply the formula for determining friction loss in 2½"-hose (FL = 2Q squared) to the previously established limit of 15 PSI per 100 feet, the maximum flow that a 1,000-foot line could handle would be between 250 GPM and 275 GPM. To allow for minor variations in elevation between pumpers, a good operating practice is to consider a 2½"-supply line as capable of supplying a maximum of 250 GPM.

In your determinations of the maximum flow to be expected through a 2½"-supply line, consider the types of nozzles commonly used with 2½"-attack lines. For example, a typical adjustable gallonage 2½"-nozzle varies from 150 GPM to 250 GPM. Since the rate of flow is controlled by the firefighter at the nozzle, it is impossible for the pump operator, without flow meters, to determine how much water is flowing at any time. The water supply officer and pump operator need to know the maximum flow setting of the nozzle and the hose layout must be configured to move enough water to supply it. Remember to consider the number of attack lines in service as an indication of the required capacity of the supply line. Also, keep in mind that one 2½"-supply line will provide only enough water for one 2½"-attack line, or two 1½"-attack lines, so if additional attack lines are required, additional supply lines also will be needed.

Figure 4-1. Above, left, Relay operation with single line of 2½"-hose.
Figure 4-2. Above, right, Relay operation with dual line of 2½"-hose.

As a general rule, a relay operation that uses only one 2½"-hoseline is an exercise in futility because the very limited capacity of a single 2½"-supply line will not support an effective attack on anything but a very small fire. If 2½"-hose is all that is available, dropping multiple lines on the initial layout should be standard practice. An alternative is for first-arriving mutual aid engine companies to lay parallel hoselines immediately upon arrival to support the increased attack capabilities of the additional engine company personnel. To do this, your pumpers must have split hose beds to allow laying two lines at the same time. By connecting two hose beds, you have the option of laying one long line or two shorter ones. Even with multiple lines, however, 2½"-hose will not support the large-scale attacks that a good fire department is capable of making. For this reason, progressive fire departments have eliminated 2½"-hose as supply line and substituted 3"- or larger hose.

THREE-INCH HOSE

If your fire department has adopted 3″-hose with 2½″-couplings as the standard for supply lines, you can use the simplified formula to determine friction loss (FL = Q squared) to the 15 PSI per 100 feet predetermined limit. Based on this, 3″-hose would be able to handle nearly 400 GPM, enough to supply a 1½″-attack line in addition to a 2½″- or up to three 1½″-lines. If a pumper has a split hose bed, an engine company can lay a dual line of 3″-hose on its initial layout. A dual line of 3″-hose can carry the full capacity of a 750-GPM pumper, and can move 1,000 GPM approximately 600 feet without exceeding 150 PSI in friction loss.

FIREGROUND HYDRAULIC CALCULATIONS

All friction loss calculations are based on the amount of loss in a 100-foot section of hose. The amount of friction loss in a section of fire hose varies with the square of the change in flow. For example, if flow is doubled, the resulting friction loss will be four times greater. If dual lines of the same size and length are used, total friction loss in the hoselines will be one fourth the loss in a single line because the flow will be divided; only half as much water will be transported through each line.

Three-inch hose provides a good basis for simplified fireground hydraulics calculations. Pump operators and water supply officers at working fires need simple methods, not complex mathematical formulae. One approach that works well is to use the "rule of squares."

To use the rule of squares, multiply the first number of the GPM (the number of hundreds of gallons) by itself to get a good approximation of the friction loss in a 100-foot section of 3″-hose. For example, 300 GPM would generate 9 PSI per 100 (3 × 3) friction loss; 400 GPM would create 16 PSI per 100 (4 × 4) friction loss.

The "hand method" (Figure 4-3) is an easy way to calculate friction loss in 3″-hose. Starting with your thumb, each finger is marked with the numbers one through five, representing the desired flow in hundreds of GPM. Again starting with the thumb, the top of each finger represents the numbers one through five. The top number is used as the constant to calculate the friction loss for various flow rates. For example, to determine friction loss at 300 GPM, multiply the two numbers on the middle finger together, i.e., 3 × 3 = 9. Your friction loss is 9 PSI per 100. The hand method also makes it easy to estimate friction loss

Figure 4-3. Hand method used for friction loss calculations up to 500 GPM through 3″-hose with 2½″-couplings.

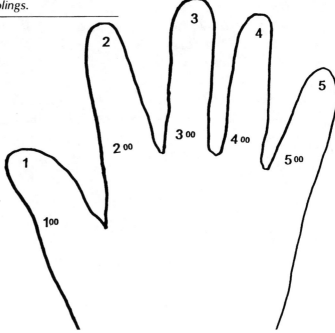

Figure 4-4. Hand method used for friction loss calculations above 500 GPM through 3″-hose with 2½″-couplings.

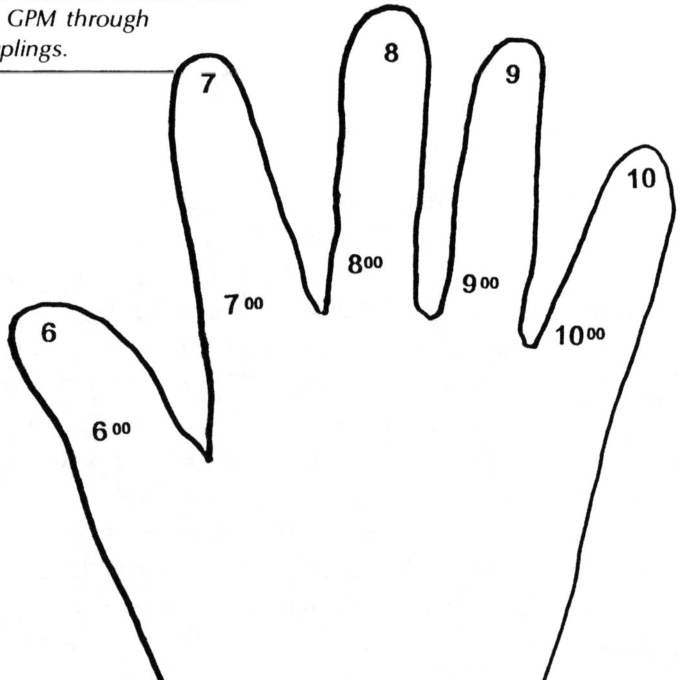

when the flow rate is not an even 100 by simply crossmultiplying between adjacent fingers. If you have 350 GPM, you would multiply the bottom 3 times the top 4 to get a friction loss of 12 PSI per 100. You can determine friction loss for larger flows by using your second hand with the numbers six through ten for flows from 600 to 1,000 GPM, and the constants from six through ten. Figure 4-4 shows the extended hand method for higher flows.

The hand method can be used again in a two-step process for determining the friction loss in other hose sizes. To do so, you apply a predetermined factor to the 3″-figures. For example, the friction loss in 2½″-hose is approximately twice that of 3″-hose. The first step to determine the friction loss in 2½″-hose is to determine the friction loss in a section of 3″-hose. Then, multiply the results of the 3″-calculation by two. With the two-step process, 200 GPM would generate 4 PSI per 100 (2 × 2) loss in 3″-hose. If you double this, 200 GPM would result in approximately 8 PSI per 100 feet flowing through 2½″-hose. You can convert the 3″-hand method results to other sizes by using the following factors:

Table 4.6. Friction Loss Conversion Factors From Three-Inch Hand Method

Size of the Hose	Conversion Factor
1½″-Line	3″-FL × 26
1¾″-Line	3″-FL × 12
2″-Line	3″-FL × 6
2½′-Line	3″-FL × 2
4″-Line	3″-FL / 5
5″-Line	3″-FL / 15
6″-Line	3″-FL / 30

If we want to find out how much friction loss a 100-foot section of 4″-hose would generate with 1,000-GPM flowing, we first solve for 3″-hose with a 1,000 GPM flow. By multiplying 10 × 10, the friction loss in 3″-hose would be 100 PSI. For 4″-hose, divide by five. The friction loss in 4″-hose would be 20 PSI per 100-foot section.

LARGE DIAMETER HOSE

The NFPA considers any hose with an inside diameter of 3½" or larger to be large diameter hose. Four- and 5"-hose are the most common sizes used in the fire service, but due to differences in construction, friction loss in large diameter hose varies greatly from one manufacturer to another. Because there are significant differences among single-jacketed, double-jacketed, and rubber-coated hose, friction loss calculations are not very precise. However, the factors from the conversion table give some idea as to the relative friction loss that can be expected in hose of different sizes.

A common misconception is that friction loss in large diameter hose is negligible and not worth considering. Many fire department personnel believe that because their pumpers carry LDH, friction loss is no longer a problem. This is not true, especially for 4" or smaller hose. If you use the standard for 2½- and 3"-supply lines, limiting the flow to produce a maximum of 15 PSI/100 friction loss, 4"-hose will move approximately 800 GPM, nearly the same capability as a dual line of 3"-hose. Four-inch hose is not large enough to move the rated capacity of a 1,000-GPM pumper through a 1,000-foot supply line. On the other hand, it will supply enough water to mount a massive attack on a fire within 1,000 feet of a source with only a single supply line and no intermediate relay pumpers. Experiments have shown that an attack pumper with 1,000 feet of 4"-hose used in a direct lay from a hydrant with residual pressure of 40 PSI will supply a master stream flowing 350 GPM. If you add a pumper at the hydrant (using a four-way hydrant valve) and increase the pressure to 150 PSI, the attack pumper will be able to flow 850 GPM through the master stream device. Although 4"-hose will provide enough water for an effective initial attack before an engine is connected into the line at the hydrant, you need 5- or 6"-hose to realize the full benefit of large diameter hose.

If you apply the 15 PSI/100 limit to 5"-hose, it would be possible to move 1,500 GPM a distance of 1,000 feet through a single line. This would handle the full capacity of most of the pumpers commonly used in the fire service, and would provide all of the water a single engine company could possibly apply to a fire. A greater advantage is that a single pumper could supply 1,000 GPM to a fireground emergency more than 2,000 feet from the source through a single line of 5"-hose. This not only provides a good water supply, but simplifies the operation and, by eliminating the need for a relay pumper, increases the water supply's reliability.

One of the biggest advantages of 5″-hose is that it eliminates the need for a supply pumper at the fire hydrant to supply the attack pumper in most cases. If a hydrant has 40 PSI residual pressure with 1,000 GPM flowing, the attack engine could be supplied through nearly 700 feet of 5″-hose by hydrant pressure alone. Even at fireground locations 1,000 feet from a hydrant, an attack engine should flow 800 GPM through a single 5″-line without a separate supply engine. Figure 4-5 is a photograph of a pumper flowing 800 GPM with the residual pressure at the hydrant forcing the water through a single line of 5″-hose.

Using 5- or 6″-hose in areas with a strong hydrant system allows the fire department to reduce its response level by one pumper on fires that are within 1,000 feet of a hydrant. This reduces the number of pumpers needed, and more importantly, frees the firefighter who would have had to serve as a pump operator at the hydrant, to assist in the firefighting. When engine company manning levels require extra personnel for fire attack who would have responded on the water supply pumper, their engine can be used as a reserve pumper on the scene. This frees the personnel to assist immediately with firefighting. An alternative is for the second engine to lay a line to another hydrant, providing more diversity in the water supply and greater reliability. With chronic personnel shortages in both volunteer and paid departments, releasing the pump operator to perform other duties on each response is a significant advantage.

Large diameter hose is particularly useful in cities whose hydrant systems cannot provide the needed fire flow in all areas. Because LDH can move large quantities of water over long distances, strong hydrants some distance away can be used to supplement the flow from weak hydrants. When alternate sources such as rivers, lakes, or other drafting supply points are available, large diameter hose can be the means to supplement the hydrant system in emergencies.

The minimal friction loss in large diameter hose means that pumpers can operate at much lower net pump pressures. The maximum rated capacity of a pumper decreases as the net pressure increases. A standard pumper is rated to supply 50% of its capacity at a net pump pressure of 250 PSI, 70% at 200 PSI, and 100% when the pressure does not exceed 150 PSI. At still lower net pump pressures, pumpers can supply more than their rated capacity. This capability is especially important in rural areas where tankers are loaded from draft. Field tests have proved that using dual hard suction hoses on the intake and large diameter hose on the discharge, 1,000-GPM pumpers have been able to load tankers at a rate of more than 2,000 GPM. A high flow rate when filling tankers is critical for operating a high-capacity water shuttle.

Large diameter hose also is useful to fill tankers from a hydrant. If you connect the large diameter hose to a 4½″-hydrant outlet, it is possible to load tankers at more accessible locations some distance from the hydrant without appreciably reducing the fill rate below the flow that could be expected from a soft suction hose.

Large diameter hose has two problems that limit its use in the fire service. First, the weight of the hose when filled makes it too difficult to handle on the fireground. This is a valid concern. Table 4.7 compares the weights of different types of commonly used supply lines when they are full of water.

Table 4.7. Weight of a 100-foot Length of Fire Hose When Full of Water

Hose Size	Capacity in Gallons	Weight Water	Weight Hose	Total
2½″	25.5	212	100	342
3″ (2½″-cplg)	36.8	307	130	471
4″	65.25	546	70	616
5″	102.0	853	100	953

NOTE: Since most LDH is made of lightweight synthetic material with sexless aluminum couplings, the average 100-foot section weighs less than a typical 100-foot section of 2½″- or 3″-hose.

This chart shows that LDH is virtually immovable when filled. Make sure your supply line is out of the way and not blocking access to the fireground before it is initially charged. This is a good practice in any situation, but essential with LDH.

A second, equally valid concern is that a 1,000-foot line of 5″-hose requires more than 1,000 gallons of water for initial filling, a potential problem at rural locations. Remember, however, that while it takes a large volume of water to fill the hose initially, all water supplied to it from that point on passes through it. Transporting water with LDH requires a source no larger than smaller lines would to supply the same amount of water to the incident scene. The reduced friction loss, along with the ability to move high volumes of water over long distances at low discharge pressures, will more than compensate for the amount of water required to fill LDH initially.

Figure 4-5. Large Diameter Hose stored on hose reel on pumper.

All things considered, large diameter hose can be a plus for any fire department, urban, suburban, or rural. From an operating standpoint, 4″-hose does not have the water transport capabilities of 5- or 6″-hose, nor will it do everything LDH can. On the other hand, except in specialized applications, the difference in cost of 4″- versus 5″-hose is not great enough to justify purchasing the smaller size.

A WATER DELIVERY SYSTEM

As high-horsepower diesel engines have gained favor in the fire service, the capacity of the pumps on new fire apparatus has increased steadily. It is rare now for a fire department to purchase a 750-GPM pump on a new piece of apparatus; pump capacities of 1,250 and 1,500 GPM are common. Although fire pumps are important, they are just one element of a water delivery system. To take advantage of their capacity, fire departments must have sufficient water flow devices, and enough properly sized hose to deliver the water to them. On the supply side of the pump, there must be enough large hose to supply the water to support these attack lines. Many pumpers carry 2½″- or 3″-hose for supply lines. It simply is not cost effective to purchase a pumper capable of massive fire attacks without purchasing the correct hose to go with it. One such fire department has four 1,500-GPM pumpers, all loaded with 2½"-hose. It would take six parallel lines of 2½"-hose to supply just one of these pumpers over a 1,000-foot distance. It does not make sense to buy a 1,500-GPM pumper and then load it with 250- or 400-GPM hose. With five-inch hose, a pumper's full capacity can be delivered to the fireground without laying any additional lines. If you equip a pumper with hose lines sized to supply its maximum capacity with a single supply line, you need not worry about deciding how many hoselines will be used. This is one less decision for the incident commander to make. Also, rest assured that your fire attack will not be affected by the inability of the water supply pumper to transport enough water to the attack pumper.

5. Testing Water Supply Apparatus

THERE ARE FOUR CATEGORIES of apparatus that constitute one of the most important parts of a fire department water delivery system. Each type of apparatus should be tested according to the function it was designed to carry out.

ATTACK PUMPERS are designed and constructed to apply water to a fire. They must be able to carry enough fire hose, including preconnected lines, to support normally anticipated attacks.

WATER SUPPLY PUMPERS are designed and constructed to supply water from a source to an attack pumper at the incident scene. They should carry enough fire hose to deliver rated capacity to a point at least 1,000 feet from the source, and enough specialized equipment to operate efficiently under a wide range of conditions.

MOBILE WATER SUPPLY APPARATUS UNITS are designed to transport water between the supply source and the incident scene. The emphasis is on water hauling ability; the units should be designed to fit the needs of the area where they operate.

COMBINATION UNITS are designed to combine two or more of these specialized functions in one vehicle.

GENERAL EVALUATION

When you evaluate any of the types of apparatus defined above, ask first if it is designed to do its anticipated job; and, second, if it can perform well while moving water.

The best way to answer these questions is to conduct practical "hands-on" tests under conditions designed to expose any deficiencies that need to be corrected.

ATTACK PUMPERS

To evaluate an attack pumper, first consider whether it can make an initial attack on a fire in a timely manner. Can it apply enough water to control a fire in its early stages? Next, can it support the sort of massive defensive attack for a fully involved building, and sustain that level of attack for extended periods of time? Finally, is it possible to operate this unit with a minimum number of people?

Initial Attack

Check initial attack capabilities of the pumper.

You typically launch your initial attack with the water carried on the apparatus. With preconnected attack lines an engine company can make the attack quickly, and with a minimal effort, so the attack pumper should carry a variety of sizes and lengths of preconnected hose. It also must carry all of the equipment and tools needed for structural firefighting, and enough firefighters to use them, without having to wait for additional apparatus to arrive. If a good pressurized water system is available, you would want the attack pumper to have enough hose to lay a supply line from the closest hydrant while it is approaching the incident scene. This way the engine company can supplement the water it carries, and make the transition to an external supply, without interrupting the flow of water to the attack lines.

Figure 5-1. Pumper Tanker equipped with 1,250 GPM pump, 2,000 gallon water tank, 10'' gravity dump, and numerous preconnected attack lines.

Initial Water Supply

Weigh the apparatus.

It is not always true that the more water an attack pumper carries, the better. With the extra weight of a larger tank, and the 8.35 pounds per gallon that water weighs, each gallon you add to a pumper adds approximately 10 pounds to the vehicle's gross weight. A tank constructed of lightweight material reduces overall weight somewhat, but your absolute minimum would be 9 pounds per gallon; a figure of 9.5 is more realistic. A pumper that carries 1,500 gallons of water weighs approximately five tons more than one with 500 gallons, based on the additional water and the tank to carry it. In actual practice, the difference is greater. To carry this added weight requires a heavier chassis and drive train, which add to the vehicle's gross weight. Then, you face the risk that this type of apparatus will be overloaded when it is put into service. Even a vehicle that is within its permissible weight limits when it leaves the factory can exceed its Gross Vehicle Weight Rating (GVWR) when equipment is put into the compartments; this can make it unsafe for use in responding to an emergency. Weigh each pumper, fully equipped, and with a full crew aboard, to check its GVWR.

In general, attack pumpers can carry only up to 1,000 gallons of water before they need a tandem axle chassis and a heavy-duty drive train. From an operational efficiency standpoint, the physical size of a large water tank can make it difficult for a pumper to carry enough equipment for effective fireground operations. It could be that the hose bed, and many of the compartments, are so high off the ground that laying out the hoselines and reloading them, as well as gaining access to needed equipment, become difficult and time-consuming. Some pumpers with high-capacity water tanks (1,000 gallons or larger) have smaller hose beds, and frequently cannot carry enough hose to move as much water as the pump can transport more than a distance of 1,000 feet. If your equipment is inaccessible or difficult to reach, you may need additional help with initial attack, and water will not reach the fire as quickly.

Measure the maximum flow from the tank to the pump.
Many fire department pumpers cannot get water from the tank to the pump at flow rates high enough to supply the pump's rated capacity. NFPA 1901, *Standard for Pumper Fire Apparatus*, specifies a minimum flow rate of 500 GPM from the tank to pump, but gives fire departments the option of specifying a higher maximum flow rate. Unfortunately, fire departments have not taken advantage of this opportunity. Some fire pumps, especially those on older pumpers constructed when requirements were less stringent, have restrictive plumbing that limits the maximum flow from the tank to less than 300 GPM. More alarming is that many pump operators are not aware of this limitation and frequently try to exceed maximum available flow, causing cavitation in the pump and fluctuating pressure on the attack lines. Tests of attack pumpers should include flow tests through the normal piping from the tank to the pump.

PRECONNECTED ATTACK LINES

Determine the optimum pressure for preconnected lines.
While nearly all attack pumpers carry preconnected hoselines, their length, size, and location vary greatly. This makes sense because each fire department operates under different conditions and faces different types of fire problems. You can change the makeup of the preconnected lines on each pumper from time to time, but each should have a standard length, type of hose, and nozzle for normal operations. Variable flow nozzles give your fire attack teams more options for adjusting the amount of water they use, but these nozzles complicate the pump operator's job. A standard practice is to leave the adjustment of a variable flow nozzle on a predetermined GPM setting when it is stored in the hose bed, and then base the standard operating pressure for each line on the length of the hose and the nozzle's normal setting. The data in Table 4.3 in Chapter 4 show the effect that changing the flow

setting on a variable nozzle has on nozzle pressure, if the engine pressure is not adjusted to compensate for the different amount of friction loss in the line. The numbers suggest that using 1¾"-hose to supply 1½"-nozzles, and 2½"-hose for 2½"-nozzles, limits variations in nozzle pressure, over the range of flow these nozzles can supply, to maintain effective fire streams without changing engine pressure. Friction loss calculations are helpful in determining the capabilities of a preconnected line, but you should base standard operating pressure on flow tests of each line as it is normally carried.

Individual discharge pressure gauges are essential to maintain the correct operating pressure for each attack line when more than one is in use. And, unless these gauges are accurate, their value is limited. Test them periodically, and calibrate them accurately. Liquid-filled gauges typically are not adjustable in the field; they must be repaired by the manufacturer or replaced when they go out of calibration.

TESTING GAUGES

<u>Calibrate all gauges.</u>

Use a calibrated test gauge to test the master pressure gauge on each pumper. Then calibrate the master gauge and test individual line gauges by opening the discharge valves with the pump under pressure and the outlets capped. After the air in the line between the discharge manifold of the pump and the individual outlet escapes through the bleeder valve, the line gauge reading and master pressure gauge reading should match. Significant variations are a warning to calibrate, repair, or replace the gauge. To quickly check the calibration of all apparatus gauges, have the operator note that each reads zero when the pump is dry. Zero calibrating is a simple process; try to have operators perform it as part of a pump's routine operational check. Calibrating the compound gauge on the pump's intake side is especially important because the water system could be damaged if incoming pressure drops below 0 when it is operating from a hydrant. A compound gauge that reads 20 when it should read 0 could force the pump to cause a vacuum on the water system, and go into cavitation, all while the gauge shows adequate pressure still in the system.

The vacuum reading on the compound gauge is the first sign that you are nearing maximum flow when operating from the tank, water is being discharged faster than it can flow into the pump and there is a danger of cavitation. It is also a way to estimate lift when operating from draft, and warns the pump operator that maximum flow is imminent before the pump begins to cavitate. It is difficult to read accurate vacuum measurements because the movement of the pointer from 0 to 30 inches of vacuum covers such a small portion of the scale. Beware: even minor errors in adjusting the zero calibration of the compound gauge can allow the pump to go into cavitation before an operator realizes there is a problem.

FLOW TESTS

<u>Verify that the pump can supply all preconnected attack lines from the tank.</u>

With variable flow nozzles, it is the nozzle setting that controls amount of water moving through the fire pump, not the pump operator. Since fire attack takes place away from the apparatus, and actual water flow is not controlled by the pump operator, the pump must be able to supply as much water as the preconnected lines demand. Run tests each time you change the number or type of preconnected lines on the apparatus to make sure the fire pump can supply maximum required flow.

The best way to test an attack pumper's flow capabilities is to put the pump into service, use water from the tank on the apparatus, and flow all preconnected lines simultaneously with nozzles at their maximum settings. Apply the standard operating pressure to each line, and observe the intake gauge to make sure that the pump does not go into cavitation when all the lines are flowing at maximum rate. If the pump cannot supply all preconnected lines simultaneously, it is safest to eliminate some of them and mark the flow limitations clearly on the pump panel. Caution all operators to limit the number of preconnected lines put into service before the external water supply is established.

MASTER STREAM TESTING

<u>Determine the best operating preassure to supply the master stream device.</u>

Preconnected master stream devices are used to make a massive fire attack when you need a large volume of water, but have minimum staffing for the operation. Many of these devices can be removed for portable use or supplied through a built-in connection while mounted on the apparatus. If you need to use the master stream mounted on the apparatus, you will not be able to achieve the desired flow if the tank to pump line is too small. When you purchase the unit, specify a maximum flow rate from the tank to the pump that will accommodate the maximum flow from the master stream device.

When a master stream device is to be supplied by an internal waterway on the apparatus, it should be large enough to maintain desired nozzle pressure at its maximum flow setting without exceeding 150 PSI discharge pressure on the pump. This is the maximum net pump pressure the pump is required to produce at its rated capacity. Test this capability when you first install the device, and check it each time you change the valves, piping, or the device itself. With these tests, you can determine the proper operating pressure for supplying the master stream under various operating conditions. Mark this pressure clearly on the pump control panel.

Table 4.5 (Chapter 4) gives estimates of expected maximum flows from 2½″- or 3″-lines used as attack lines to supply water to a master stream device. It takes more than one of these hoselines to supply enough water for the master stream nozzle to discharge its maximum flow when it is removed from the apparatus and used elsewhere. Using LDH to supply the master stream eliminates the need for multiple lines, but makes the device much more difficult to move while it is in service. Along with being less mobile, a portable master stream device also is less stable with one line of large diameter hose than with multiple lines brought into the inlet from different directions.

SUPPLYING THE MASTER STREAM

Measure the maximum flow from the preconnected master stream device.

A 500-gallon capacity or less apparatus tank can be used to supply the master stream only in special situations, for example, when you need a heavy flow for a very short time to dislodge an inaccessible beam or to knock down a fire-damaged chimney. Apparatus with larger tanks, say, a pumper tanker that carries 2,000 gallons or more, could use the master stream to launch a massive initial attack from the tank as soon as you arrive at the scene. In applications of these types, maximum flow rate from tank to pump is critical, so you should test the fire pump's ability to supply the maximum flow the master stream device requires from the tank on the apparatus when it is originally installed and when any changes are made. Do this by setting the master stream device to the maximum flow position with the standard operating pressure on the pump, and observe the compound gauge on the intake to see whether there is a vacuum reading; if so, there is a danger of the pump going into cavitation. When you must limit flow to the master stream device to prevent damaging the pump, mark the maximum setting clearly on the pump panel, and in a conspicuous place on the device itself, preferably near the point where the adjustment will be made.

WATER SUPPLY PUMPERS

Test the pump regularly and maintain a permanent service record.

A pumper intended for use as a water supply pumper must have a pump that can supply its full rated capacity under a variety of conditions. Verify the pump's capability by conducting a service test at least annually and recording the results in the pump's permanent record. Note any deterioration in its performance. Figure 5-2 is an example of the type of pump test log sheet you can use to record annual service test results. Use the results to schedule major pump repairs and consider the condition of the pump when you set apparatus replacement priorities.

CONDUCTING A PUMP SERVICE TEST

Perform a service test annually.

When a pumper is constructed, it is the Underwriters Laboratories that supervises the series of initial tests that ensure that the apparatus meets all of the requirements of the Insurance Services Office (ISO) and NFPA 1901. These test results are included with the manufacturer's certification when the apparatus is delivered. They should be made part of the unit's permanent record. The test information also is recorded on an Underwriters' test plate, usually displayed near the unit's pump control panel. In your annual service tests, evaluate the condition of the pump and apparatus by trying to duplicate, as nearly as possible, the conditions when the initial tests were made, and compare the results to the data on the Underwriters' test plate. Make all tests with the pumper operating at draft with no more than a 10-foot lift. Table 5.1 gives the requirements for each part of the test.

Testing Water Supply Apparatus 91

FIRE DEPARTMENT: DATE;
LOCATION: PUMPER:
TESTED BY: CAPACITY:

TIME	PITOT READING	TIP SIZE	FLOW IN GPM	GAUGE READINGS ON PUMPER			
				PUMP PRESS	ENG RPM	OIL PRESS	ENG TEMP
REQUIRED 100% CAPACITY							
REQUIRED 70% CAPACITY							
REQUIRED 50% CAPACITY							

PRIMING TIME_____ PACKING_____ VIBRATION _____
NOTES:

Figure 5-2. Pump Test Worksheet for recording readings during annual service tests.

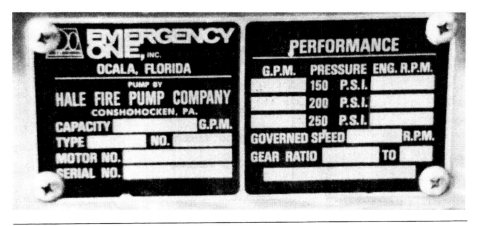

Figure 5-3. Underwriters' test plate for recording results of certification tests of the pump when the apparatus is constructed.

Table 5.1. Service Test Requirements

Gallons per Minute	Net Pump Pressure	Length of Test
100% of rated capacity	150 PSI	20 minutes
70% of rated capacity	200 PSI	10 minutes
50% of rated capacity	250 PSI	10 minutes

NET PUMP PRESSURE

Calculate expected net pump pressure.

Net pump pressure is a measure of the total pressure a fire pump develops, or, because a pump operating from draft during service tests has a vacuum (actually a negative pressure less than atmospheric pressure) on the intake side of the pump while the discharge is operating at a positive pressure, the net pump pressure actually is the algebraic sum of the negative pressure on the intake and the positive pressure on the discharge side. The most accurate way to determine a pump's net

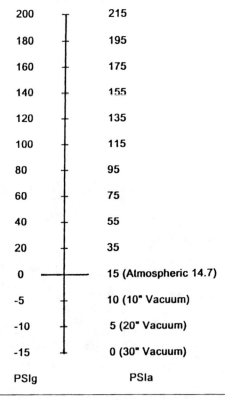

Figure 5-4. Chart showing relationship of positive and negative pressure.

pressure is to convert the intake pressure from inches of vacuum to pounds per square inch of negative pressure. A reading of 1" of mercury equates to a negative pressure of approximately .5 PSI, equivalent to about one foot of lift. You could then add the negative pressure on the intake directly to the discharge pressure to determine net pump pressure. For example, assume a fire pump is operating from draft with a discharge pressure of 150 PSI. The compound gauge registers 20 inches of vacuum on the intake. Convert the vacuum reading to an approximate pressure of 10 PSIg below atmospheric, and, by using a scale like the one shown in Figure 5-4. So, starting 10 PSI below atmospheric at an absolute pressure of 5 PSIa, and developing a pressure of 150 PSIg, would give an absolute pressure of 165 PSIa, and indicate a net pump pressure of 160 PSI. (-10 PSIg + 150 PSIg = 160 or 165 PSIa-5 PSIa = 160 PSI).

Few compound gauges have scales detailed enough to give accurate vacuum readings, so the most practical way to determine the net pump pressure is by using conversion tables. NFPA 1901 includes a table of allowances for friction loss in suction hose, and a formula for determining a correction factor to use in calculating net pump pressure.

According to NFPA 1901, large fire pumps need large hard suction hose as shown in Table 5.2.

Table 5.2.

Pump Capacity	Suction Hose. Size	Suction Hose. Number	Maximum. Lift
750 GPM	4½"	1	10'
1,000 GPM	5"	1	10'
1,250 GPM	6"	1	10'
1,500 GPM	6"	1	8'
1,500 GPM	6"	2	10'
1,750 GPM	6"	2	8'
2,000 GPM	6"	2	6'

ESTIMATING NET PUMP PRESSURE

The formula listed in NFPA 1901 for determining the correction factor is based on the length of the suction hose and the lift in feet. The test layout for service tests specifies that no more than two sections of suction hose can be used and that lift cannot exceed 10 feet. With a pump that has a capacity of 1,500 GPM or more, the lift may have to be less than 10 feet as listed in Table 5.2. If you observe these limitations, and use a standard correction factor, you can avoid complex mathematical computations.

A pumper drafting through 20 feet of the appropriate size of suction hose from a 10-foot lift measures between 7 and 9 PSI (16 to 21 inches of vacuum) of negative pressure when the pump is flowing 100% of rated capacity. This measurement changes by 1 PSI whenever the lift changes by 2.3 feet. For example, for a lift of 12 feet versus 10 feet, the correction factor would be 8 to 10 PSI. At 8 feet, the correction factor would be only 6 to 8 PSI. Adding a 10-foot section of suction hose raises the correction factor by 1.5 PSI at 100% of rated capacity. Given that a fire department pumper is required only to be able to develop 22 inches of vacuum (less than 10 PSI), you can see that many pumpers would be unable to supply 100% of rated capacity when lift exceeds 10 feet, or more than two sections of suction hose are required.

When flow through a pump is less than 100% of capacity, the amount of friction loss in the suction hose is reduced. The correction factor for determining pump pressure is lowered by 1 PSI for the 70% test, and 2 PSI for the 50% test.

It is a common practice to ignore the correction factor during annual service tests and to set the pump to operate at the specified test pressure. In this case, net pump pressure is approximately 5% higher than the pressure gauge indicates and exceeds the original test conditions. Because all fire pumps must meet a 10% overload test before they leave the factory, a pumper in good condition can do this, but at a higher expected engine RPM than is shown on the UL certification test plate. If you conduct the tests under the same conditions each time, you can monitor the condition of the fire pump and detect deterioration as it occurs. For more accurate tests, use the correction factor when they are made.

TESTING LAYOUT

The most accurate way to conduct a service test is to use a smooth bore nozzle with an assortment of tips, or a pump test kit with inserts of various sizes, and a Pitot gauge, with the pickup tube inserted into the stream of water, and the opening perpendicular to the direction of movement, to determine the flow. Use the formula

$$29.7 \times D^2 \sqrt{NP} = GPM$$

where

GPM = Flow from the nozzle
D = Diameter of the nozzle opening
NP = Pressure reading on the Pitot gauge

For a smooth bore monitor nozzle operating with a nozzle pressure of 80 PSI, and a 1½″-tip on the nozzle; the calculation is

$$29.7 \times (1.5 \times 1.5) \times \sqrt{80} = 597.7$$

It is difficult, if not impossible, to conduct a service test using a directly connected master stream device permanently mounted on the engine. For a 50% capacity test, the net pump pressure must be 250 PSI, and the pressure at the nozzle should be less than 100 PSI to get an accurate Pitot gauge reading. There must be enough friction loss between the pump outlet and the master stream device to meet both requirements. It is possible to partially close the discharge valves to adjust friction loss in the lines, but the obstruction generates too much turbulence in the stream, because of the short length of the pipe from the pump to the inlet of the master stream device, to get an accurate Pitot gauge measurement. You generally need at least 100 feet of hose for the turbulence to dissipate enough to measure nozzle pressure accurately with a Pitot gauge. Hoselines must be large enough to deliver the pump's full rated capacity, with the required nozzle pressure, during maximum flow tests. A usable layout is shown in Figure 5-5, and Table 5.3 lists nozzle tip sizes, number of 2½″- or 3″-hoselines, and desired pressure readings for 750-, 1,000-, and 1,250-GPM capacity pumps.

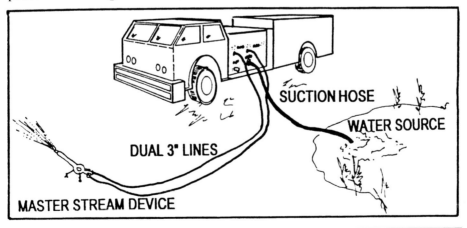

Figure 5-5. Diagram of a typical test layout for conducting annual service tests on a pumper.

96 Fire Department Water Supply Handbook

Figure 5-6. Conducting an annual service test on a pumper.

Table 5.3. Service Test Layout and Requirements

GPM Flow	Nozzle Size	Nozzle Pressure	Discharge Pressure	Net Pump Pressure	Hose-lines
750-GPM Pump					
375	1¼''	66	244	250	1
525	1½''	62	193	200	2
750	1¾''	68	142	150	2
1,000-GPM Pump					
500	1½''	57	244	250	2
700	1¾''	60	193	200	2
1,000	2''	71	142	150	3
1,250-GPM Pump					
625	1¾''	47	244	250	2
875	2''	54	193	200	2
1,250	2¼''	69	142	150	3

NOTE: Pumpers larger than 1,250 GPM generally need multiple flow devices, with the individual flows added together to determine total flow.

TESTING PROCEDURE

Discharge pressure in a centrifugal fire pump is a function of impeller speed, incoming pressure, and amount of water passing through the pump. In general, with a given engine speed and an adequate supply with constant residual pressure coming into the pump, the more water moving, the lower the discharge pressure. When the flow rate is less, the resistance of the water's movement creates a "back pressure" and the pressure generated by the pump increases. To conduct a valid service test, a target flow rate has to be met while a specified discharge pressure is maintained at the pump. Set the flow rate, as determined by measuring nozzle pressure, by partially closing the discharge valves that supply the master stream device, while maintaining the proper test pressure at the pump by adjusting the engine speed with the throttle. Each adjustment affects the other. If you open the discharge valve further to increase flow, the increased flow causes the discharge pressure to decrease. If you increase the throttle to restore the pump discharge to the specified test pressure, the nozzle pressure also increases, with a corresponding change in flow. Establish target test conditions by coordinating the following adjustments.

1. Position the apparatus for drafting and immobilize it. Lift should be as close as possible to 10 feet from the surface of the water, not the depth of the strainer, to the intake of the pump.
2. Connect 20 feet of hard suction hose equipped with a floating strainer to the pump inlet. If no floating strainer is available, use a barrel strainer, but submerge it at least 18" under the surface of the water to reach the full rated capacity of the pump without causing a whirlpool to form.
3. Connect the hose and appliances as specified in Table 5.3 to make the 100% capacity test. Engage the pump and set all of the controls to the proper position for this test.
4. Operate the priming mechanism. Record the time required to fully prime the pump and compare that time with your records to check priming efficiency. You should be able to establish draft in no more than 30 seconds; a typical pump will complete priming through 20 feet of suction hose within 10 to 15 seconds.
5. Open the discharge lines and adjust the throttle to bring the engine to the same RPM recorded on the UL test plate for the 100% capacity test. While you use the throttle to maintain the specified RPM, adjust the discharge valves to bring the pressure reading at the pump to 142 PSI, allowing 8 PSI for the lift and friction loss in the suction hose.
6. Measure the nozzle pressure with a Pitot gauge. Adjust the discharge valves and the throttle until the nozzle pressure and

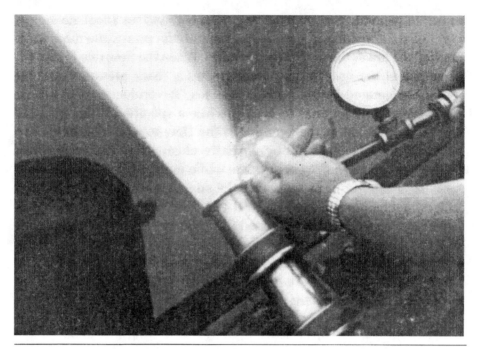

Figure 5-7. Measuring the flow with a Pitot gauge during a pump test.

Figure 5-8. Flow test kit with Pitot gauge and various size tips for testing fire pumps or fire hydrants.

Figure 5-9. Measuring the flow during a service test with flow test kit instead of hand-held Pitot gauge.

pump pressure specified in Table 5.3 are achieved. Record the reading on the engine tachometer and compare it to the specified RPM on the UL test plate. Test results should be within 10% of the initial readings. If they are not, repairs may be needed.

7. After you have made all the adjustments, lock the discharge valves in position, and continue the test for 20 minutes. During the test, make measurements frequently with the Pitot gauge, with the pickup tube held in the middle of the stream, a distance equal to ½ the diameter of the tip away from the opening. Record all readings on a test worksheet. (See a sample worksheet in Figure 5-2.)

8. Immediately after you complete the 100% capacity test, make any required changes, and use the same procedure for the 70% and 50% capacity tests, reducing the allowance for lift and friction loss from 8 to 7 and 6 respectively.

9. If all of the requirements have been met, determine the pumper's reserve capacity. Use the same test setup and parameters that you used for the 100% capacity tests, and increase the throttle adjustment until the net pump pressure is 165 PSI.

When the pump was new, it could reach 165 PSI at maximum flow, which ensured that it had at least a 10% excess capacity. If it still has, it is in excellent condition. A pump pressure lower than 165 PSI is an indication that the engine has either lost some of its power, or that the pump capacity has deteriorated, and the apparatus might need service.

10. If the apparatus has a front or rear suction, repeat the 100% capacity test with the suction hose attached to this connection. Because of piping configurations, many pumpers cannot deliver their rated capacity from draft through these suction hose fittings. It is difficult to get the piping for the front or rear suction hose through the chassis and around the springs, axles, and other components of the suspension system. To do so sometimes requires many sharp angle bends and other pipe fittings. Because of this limited space, many front or rear suction intakes use only 4½"- or 5"-pipe rather than the 6"-suction line required for pumpers with rated capacities higher than 1,000 GPM. If this is the case, consider modifying the unit so that it can reach its full capacity drafting from the front or rear suction. If you cannot modify it, make sure that the water supply officer and all pump operators know about this limitation.

OPERATIONAL TESTS

Test the prime system and automatic pressure control device.

The dry prime test is the best indication of whether or not a pumper can operate effectively from draft. To perform a dry prime test, drain the pump, close all drains and valves, and cap all pump openings. Operate the priming device until the compound gauge shows at least 22 inches of vacuum. Then, disengage the primer, and shut off the engine. After the engine has been shut down for a period of five minutes, check the compound gauge; it should maintain a reading of at least 12 inches of vacuum. If the pump loses less than 2 inches of vacuum per minute, it is probably tight enough to draft water quickly and easily. A loss of 2 inches or more per minute indicates an air leak in the pump; it will be difficult, if not impossible, to operate effectively from draft. You usually can detect large leaks by the sound of air being drawn into the pump. In addition to valves not fully closed, loose caps, open drains, and worn gaskets, the failure to meet the dry prime test may be an indication that a pump needs to be serviced. It is a simple test to make because it does not require moving any water. Perform it frequently. In fact, in paid departments the pump operator should pull a dry prime on the apparatus at the beginning of each tour of duty. In volunteer departments, incorporate the dry prime test into your operational checks; have it done weekly and recorded by the person responsible for pumper operation and maintenance.

Another important apparatus accessory is the automatic pressure control device. If a pumper is to be used primarily as an attack pumper, the automatic pressure control device protects the fire crew on one line from pressure surges when changes are made in the flow of other lines the same pumper is supplying. In water supply pumpers, the device prevents excessive pressure surges when there are large changes in flow, for example, when pumpers are operating in tandem or in a relay. All of the many different types of automatic pressure control devices in use are subject to the same requirement. To make this test, operate the pump at 100% of its rated capacity with 150 PSI net pump pressure. When you close all discharge valves slowly discharge pressure cannot increase more than 30 PSI at any time. The automatic pressure control device must be able to maintain this level of control over net pump pressures that range from 90 PSI to 250 PSI. Test this device at least monthly, and service it when necessary.

INTAKE RELIEF VALVES

Check and adjust the operating point.

Many pumpers also have intake relief valves to limit pressure surges when the pump is supplied under pressure. If the relief valve's pressure adjustment control is accessible, adjust it for each operation the same way you would set the automatic relief valve on the discharge side of the pump. If the adjustment control is not accessible because of installation practices, the valve often is left at a predetermined setting. In either case, you should test this valve's adjustment and operation on a regular basis.

A common practice is to set the intake relief valve to 150 PSI and leave it there. The 1991 edition of NFPA 1901 states that the manufacturer shall preset the system to 125 PSI unless otherwise directed by the purchaser. It is preferable to set the relief valve operation point just above the highest static pressure you expect from the hydrant system in your service area. Follow these steps to test the operation of an intake relief valve:

Figure 5-10. Intake relief valve mounted below the running board on a pumper.

1. Connect a supply line from another pumper's discharge to the intake of the pumper you want to test.
2. Determine the highest static pressure in the hydrant system, and add 10 PSI to determine the desired operating pressure.
3. Adjust the supply pumper to produce the desired operating pressure at the pump's intake to be tested as indicated on the compound gauge.
4. Turn the adjustment control on the intake relief valve counterclockwise until the valve opens, and water discharges from the valve's outlet. Turn the adjustment control slowly in a clockwise direction until the valve closes, and the flow of water stops.
5. Verify the relief valve's setting by increasing the intake pressure. The valve should open and discharge water with no more than a 30-PSI increase.

If you set the intake relieve valve this way, it limits the intake pressure to a good operating range, one that will not waste water when the pump is connected to a hydrant.

SERVICE TEST RECORDS

Record the results of the annual service test on a permanent record form.
During an ISO evaluation, your department will be expected to produce three years' worth of past annual service tests. You also need to keep records for comparison purposes so that your maintenance

PUMP TEST RECORD

PUMPER NO._____ FIRE DEPARTMENT_____

DATE_____ TESTED BY_____

LIFT_____ TESTED AT_____

MAKE_____ YEAR_____ PUMP MAKE _____ CAPACITY_____

PUMP TEST RESULTS SUMMARY				
MEASUREMENTS	TEST RESULTS			
	100% CAP.	70% CAP.	50% CAP.	10% OL.
Series/Parallel Pos.				
Gallons Per Minute				
Discharge Pressure				
Suction Pressure				
Net Pump Pressure				
Engine RPM				
RPM UL Test Plate				
Difference RPM				
% Difference RPM				
Automatic Pressure Control - Max. Inc.				

PRIMING TIME_____ DRY PRIME TEST - VACUUM AFTER 10 MINUTES_____

OIL PRESSURE_____ MAX. RADIATOR TEMP._____ PACKING ADJUSTMENT_____

Figure 5-11. Worksheet for maintaining a permanent record of annual service tests of a pumper.

department can monitor pump condition and schedule repairs before problems become serious. Keeping records also helps to monitor apparatus reliability.

Figure 5-11 is a sample service test record you can use to summarize the results of annual service tests. Fill out one of these test record worksheets each time you test a pumper, and include it in the vehicle's service record history. When you conduct the test, compare the results with previous records and note any trends. If a pump needs more RPMs to develop 150 PSI of net pump pressure at 100% of its rated capacity, it is deteriorating. It is often excessive wear on the clearance rings that is the culprit, and the pump may have to be rebuilt. Pump wear is normal and a certain amount of deterioration is tolerable. With gradual deterioration, you can schedule major repairs well in advance, but if there is a noticeable difference from one year to the next, typically more than 10%, you need to investigate what could be a serious problem with the pump or associated accessories. Reduced capacity frequently is caused by the pressure relief valve's tendency to stick open and allow a portion of the water to be bypassed back into the intake.

Another common problem is an engine's inability to provide enough power to reach the required RPMs at the pump's rated capacity. You may notice a drop in oil pressure or an increase in coolant temperature after the pump has been operating for some time. Often, the first sign of engine trouble may be that the engine fails to meet the annual service test requirements or shows abnormal readings on some gauges during prolonged operation.

In any case, analyze test results thoroughly and schedule needed repairs as soon as possible.

MOBILE WATER SUPPLY UNITS

Testing parameters.

In its role in a fire department water supply system, an individual mobile water supply unit's capacity is important. Tankers transport water, so the GPM quantity each can supply is as important as the size of the water mains in a central water system.

There are a number of different ways to determine the continuous flow capabilities (CFC) of mobile water supply units. When the ISO rates a community for Fire Department Water Supply capabilities using tankers to shuttle water, it determines the capacity of each one to be credited. By testing dump time and fill time, and estimating travel time (using a constant 35 mph), the ISO decides how much water a tanker can deliver to a particular location. The rating assumes that 90% of a tanker's total capacity will be usable. The remaining 10% is attributed to incomplete filling or dumping in shuttle operations, a figure that can be too high or too low. If standardized testing procedures are used, there is no reason that a tanker's actual capacity cannot be determined as accurately as the capacity of a fire pump. Extensive field tests have shown that the amount of usable water on tankers ranges from a high of 99% to a low of 65%. The difference can be a critical factor in water supply planning.

It is impossible to evaluate a tanker accurately without knowing the amount of time needed to fill and dump a tanker. Some methods measure only the peak flow for loading and unloading, and not how the peak flow changes when the tank's water level drops as water is discharged. This approach could give you an erroneous measurement of a tanker's ability to deliver water to a fire scene.

Other factors contribute to mobile water supply vehicle efficiency. The weight handling capacity of the chassis, available engine power, and pump capacity also are important, and must be evaluated. The evaluation process not only provides an accurate estimate of each mobile water supply unit's capacity, but establishes a sound basis for improvement.

CLASSIFYING MOBILE WATER SUPPLY APPARATUS

All mobile water supply units fall into one of three general categories:

1. PUMPER-TANKER-Fire department pumper with a water tank of 1,000 gallons or more. Designed primarily as a pumper, its secondary purpose is to haul water.

Figure 5-12. Pumper-tanker equipped with a 1,000-GPM pump and a 1,000-gallon water tank.

Figure 5-13. General purpose tanker equipped with a 450-GPM pump and a 1,500-gallon water tank with a jet dump outlet.

2. GENERAL PURPOSE TANKER-Multipurpose vehicle that carries at least 1,000 gallons of water, has a booster-type pump, and carries a certain amount of equipment.
3. WATER SUPPLY TANKER-Tanker that is designed for one purpose only: to haul water. With or without a pump, it carries minimal equipment, typically only what is needed for the water supply function. Water supply tankers can be further classified as:

- WST-P   Water supply tanker equipped with a standard fire pump, 750 GPM or larger.

Figure 5-14. Water supply tanker, WST-P, equipped with a 750-GPM front-mount pump and a 2,000-gallon water tank.

Testing Water Supply Apparatus

Figure 5-15. Water supply tanker, WST-A, equipped with a 250-GPM attack pump and a 2,500-gallon water tank.

Figure 5-16. Water supply tanker, WST-T, equipped with a 400-GPM transfer pump, a 3,000-gallon water tank, and a 10''-gravity quick dump.

Figure 5-17. Trailer type water supply tanker, WST-N, carrying 6,000 gallons of water without a pump.

- WST-A   Water supply tanker with an attack pump, capable of delivering at least 250 GPM at a net pump pressure of 150 PSI.
- WST-T   Water supply tanker equipped with a transfer pump, capable of delivering at least 250 GPM at a net pump pressure of 50 PSI.
- WST-N   Water supply tanker with no pump that uses a large gravity dump for unloading or, with the appropriate external connections, permits a pumper to draft from the tank.

Determining in which category your vehicle fits is the first step in testing and evaluating a mobile water supply unit. The standards you use are dictated by the vehicle's purpose; given its design, test procedures change accordingly.

VEHICLE CONSTRUCTION

Weigh the vehicle to determine whether it is overloaded.

One measure of a mobile water supply unit's capabilities is its ability to handle its load. Somehow, during the planning and design process for mobile water supply apparatus, there is a tendency to add more water, and a larger tank, to a vehicle, even if its chassis cannot carry it. As you add gallons of water (each weighs 8.35 pounds), tank size must increase. As a general rule, for each additional gallon, a vehicle's loaded weight increases by ten pounds, so, by adding 500 gallons of water to tank capacity you increase the vehicle's loaded weight by more than two tons. Can the chassis handle this extra weight? If not, you have an unsafe vehicle with all its associated problems.

The ability of the chassis and suspension to handle the weight safely is crucial. A Gross Vehicle Weight Rating (GVWR) is assigned to each vehicle as it leaves the factory. The GVWR is based on frame strength, the capacity of the springs and other components of the suspension, and on tire size, and braking power. Exceed any of these, and your apparatus is a safety hazard. A vehicle's GVWR is limited by the lowest rating of any individual component. If you compare the fully loaded weight to the GVWR, you can determine whether it can handle a load safely. If it will not, it is time to make changes. Modifications often are needed when a commercial tank truck originally intended to haul a petroleum product is converted for fire service use. Why?

Figure 5-18. Water supply tanker, WST-N, converted from a fuel oil truck without a pump, but including a 1,350-gallon water tank equipped with a 10''-gravity dump outlet.

- First, the unit was originally designed and purchased to be a profit-making vehicle, so cost was a primary consideration in its design. It is likely that it was specified to handle its load with little or no margin of safety, not an acceptable design for an emergency vehicle.
- Second, petroleum products weigh less than water, as much as 20% less in some cases. To carry the same number of gallons of water as the fuel it was designed to haul increases the vehicle's gross weight.
- Third, the tank was designed to haul its load with a certain amount of free space to eliminate spillage. In fire service use, the normal practice is to fill it until it overflows, which, in this case, increases the fluid load by 10% or more, further overloading the vehicle.

Develop a plan to eliminate potential problems.

The easiest change you can make is to increase tire size when tires are the limiting factor. Strengthen the springs or add auxiliary springs. Replace either the front or rear axle if it is not heavy enough for the load. Reinforce the chassis if it is too light. At the difficult end of the problem spectrum is correcting brakes too small for the vehicle's weight. If you cannot upgrade them sufficiently, your only choice is to modify the tank (lower its capacity) until the vehicle reaches a safe maximum weight. If a vehicle needs extensive modifications, you need to decide whether it would be more cost effective to simply purchase a new or used vehicle than to make the changes.

Calculate the weight distribution.

Compare the amount of weight each axle carries. Weight distribution affects performance, handling and maneuverability. An ideal ratio for a two-axle unit is to have 33% of the weight carried by the front axle and the balance carried by the rear axle. A tandem-axle chassis should carry approximately 25% on the front axle and 75% on the rear. The most common problem, again predominantly on conversion units, is too much weight on the rear axle. Without enough weight on the front axle to provide good traction, steering is uncertain, and there is a tendency for the vehicle to get stuck in unpaved areas and to sway from side to side excessively during high-speed turns. Poor weight distribution typically causes one of the two axles to be overloaded, even when gross vehicle weight is within the rated capacity. Sometimes the only alternative is to block off a portion of the tank (to reduce the amount of water carried) to get the total weight on one or both of the axles within reasonable limits.

Fire apparatus usually has enough power. Most pumpers, or other types of specialized apparatus, have large power plants that give a performance level close to an automobile's. This is not true of many water haulers. Often converted from commercial tank trucks, these vehicles were purchased at the lowest workable price, and often were only marginal performers when new. The problem of underpowering occurs and is critical when water rather than a lighter petroleum-based product is hauled. If the apparatus is severely underpowered, and the condition of the rest of the vehicle warrants it, consider repowering it with a larger, more powerful engine, if the rest of the drive train can handle the increased torque safely.

Compartments can create a problem. When a vehicle is designed primarily as a water hauler, only a minimal allowance for the weight of any equipment that will be carried is included. As compartments are loaded with miscellaneous equipment to increase the vehicle's "flexibility and versatility," it is easy to cross the line to overloaded without anyone even being aware of the problem. It seems that tankers, more than any other type of apparatus, are wrecked during response and for this reason you need to know and understand a tanker's shortcomings and have your operators drive accordingly.

Measure the maximum flow rate of the pump.

A pump is not a requirement, but it can be a great asset to any mobile water supply unit, even a water supply tanker. The key is to first get an adequately sized pump and, second, to use tank to pump lines that are both long enough and properly routed. Remember that generally the maximum flow from the pump is limited by the size of the tank to pump line (Table 5.4).

Table 5.4. Maximum Flow with Various Tank to Pump Lines

Size Line	Number	Flow
2½"-Line	1	250 GPM
3"-Line	1	500 GPM
2½"-Line	2	500 GPM
4"-Line	1	1,000 GPM
3"-Line	2	1,000 GPM

In terms of line length and routing, a 90-degree bend or a tee fitting is the equivalent of many feet of pipe. Both increase friction loss significantly, and reduce maximum flow to the pump. Consider rerouting the line, increasing pipe size, or adding a second tank to pump line to reduce substantially the time needed to discharge a full load of water from a tanker through the pump.

One way to get the most out of a small-capacity pump is to add a jet assist to a large gravity dump line. A 6"-round dump outlet gives an average flow of approximately 600 GPM during unloading. You could increase the average flow from the same dump line to more than 1,100 GPM by using a PTO pump with a capacity of only 200 GPM as a jet assist to increase the velocity of the water leaving the dump outlet. This would increase the discharge rate by 500 GPM and cut the discharge time nearly in half.

FLOW TESTS

How much water can a tanker deliver to a given location? It depends on the amount of water it hauls and the length of time of each roundtrip. Although travel time **is** significant (especially when the distance between the water supply and the fire is longer than a mile), it is roughly the same for all tankers. What you must consider are how much usable water the tanker carries and how long it takes to load and unload. Flow tests can measure each of these factors.

Figure 5-19. Testing the flow rate from the pump on a tanker using a hand-held Pitot gauge.

Figure 5-20. Testing the flow rate from the pump on a tanker using a flow test kit instead of a hand-held Pitot gauge.

UNLOADING TEST

Measure the time required to unload through the pump.

If the mobile water supply apparatus has a pump, use the proper size tip or flow device and a Pitot gauge to determine maximum discharge rate. One convenient way to test flow rate is to install a stream shaper from a master stream device directly onto one of the 2½"-outlets. Choose a tip for the stream shaper that will give you a nozzle pressure that is between 30 and 90 PSI at the expected flow rate. In Figure 5-19 a tanker is being tested with a stream shaper connected directly to a pump discharge and a Pitot gauge is being used to determine flow.

You reach maximum flow rate just below the point of pump cavitation. To find the point of cavitation, put the pump in gear, open the tank to pump valve or valves, and increase the throttle until the discharge pressure will not climb higher. At that point, if the pump cannot develop 30 PSI on the Pitot gauge measuring the flow, shut down, install a smaller tip, and try again. If the pressure is still increasing when the Pitot gauge reads 90 PSI, close the discharge valve and install a larger tip on the stream shaper. Then, calculate maximum flow rate by using the same formula used to determine flow in testing the pumper:

$$29.7 \times D^2 \times \sqrt{NP} = GPM$$

It is simpler to find the flow by referring to one of the charts and tables than use tip size and measured nozzle pressure to determine discharge rates.

It is acceptable to use an electronic flow meter instead of a Pitot gauge to determine maximum flow, but Pitot gauges or pump test kits generally are more accurate because portable electronic flow meters tend to have calibration problems, and any turbulence in the discharge caused by the fittings on the discharge side of the pump can introduce inaccuracies into the measurement.

Watch the pressure gauge closely while you conduct the maximum flow test. As soon as the discharge pressure drops and the pressure gauge begins to fluctuate (indicating that the level in the tank has dropped to the point that air is being drawn into the line), close the discharge valve and shut down the pump. At this point, all usable water has been removed from the tank.

In this test, once water begins to flow do not interrupt it until the tank is empty, because if you observe the time from the opening of the discharge to the completion of the flow test carefully, you can measure the discharge time. Then, if you multiply the length of time by the measured rate of flow in GPM, you can calculate the total amount of usable water. For example, a tanker discharges its load through a 2½"-smooth bore nozzle with a 1"-tip at a nozzle pressure of 64 PSI. After six minutes of operation, the pump begins to cavitate. Calculate the usable water

$$29.7 \times D^2 \times \sqrt{NP} = GPM$$

$$29.7 \times (1 \times 1) \times \sqrt{64} = 238 \text{ GPM}$$

$$238 \text{ GPM} \times 6 \text{ Minutes} = 1{,}428 \text{ Gallons}$$

This tanker carries 1,428 gallons of usable water, an indication of how much water it could deliver to a fire scene each time it arrived during a water shuttle operation.

LOADING TEST

Determine how long it takes to fill the tanker.

After the unloading test, fill the tanker until it overflows. A direct fill line into the tank generally is the quickest way to fill it. If there is no direct fill line, but the tanker has a pump, you might be able to backfill the tank rapidly by connecting the fill line to the suction of the pump and opening the tank to pump valve. A standard fire pump probably would have a check valve in the suction line from the tank; this means that it cannot be used as a tank fill line. If there are two tank to pump lines, one probably has a check valve while the other does not. Mark the one you intend to use for filling to eliminate confusion during water shuttle operations.

You also can fill the tank through the tank fill line from the pump. When this line is 1½" in diameter or smaller, maximum fill rate is low, making fill time too long for effective water shuttle operations. Improve the fill rate somewhat by connecting the supply line to the pump's intake, and using the pump to increase the pressure to the tank fill line.

If there is no practical alternative, fill the tank through the top fill opening. This is a dangerous operation because someone must climb up on the apparatus and put a hose down inside the opening, a precarious position even if you use a special filling device to reduce the reaction, and fasten the line to the apparatus with a rope hose tool or other fastener.

When you fill at a rapid rate, watch for excessive pressure buildup inside the tank—it can cause permanent damage. When a tank fills with water, the air inside has to escape to prevent tank pressurization. Since most tanks are not designed to withstand pressure, overpressurization damages them. Adequate venting allows air to escape and prevents any appreciable pressure differential between the inside of the tank and the atmospheric pressure outside from developing. Check around the hatch opening when you fill a tank rapidly. If the hatch is closed, but not latched, it will open slightly and let air escape quickly through the opening if any pressure builds up. This is an indication that the venting is inadequate for the flow rate and you need to decrease the pressure on the supply lines to reduce flow rate. When you must have a high flow rate, and the tanker's vent is not large enough to sustain it, open the hatch before filling. This should prevent damage and minimize the effects of any shortcomings in the tank's construction.

Determine fill time by recording the time water begins to flow to fill the tank, and measuring how much time elapses until water overflows the tank. Figure 5-21 is an example of a worksheet for recording the results of this test for multiple tankers. Divide the amount of water discharged on the unloading test by the fill time to establish fill rate. If the tanker used in our previous example had 1,428 gallons of usable water and took three minutes to fill, its fill rate would be

$$\frac{1{,}428 \text{ Gallons}}{3 \text{ Minutes}} = 457 \text{ GPM}$$

CALCULATING CONTINUOUS FLOW CAPABILITY

With the information from your flow tests, you can calculate the Continuous Flow Capabilities using:

$$\frac{V}{A + 2T + B} = Q$$

where

- V = Capacity in usable water (determined in unloading test)
- A = Unloading time in test + .75 minutes for handling
- B = Loading time in test + .75 minutes for handling
- 2T = Travel time for desired location. [To rate a tanker, use 4 minutes. This equates to two miles of road distance at an average speed of 35 mph (1 mile from water or two miles roundtrip.)]

TANKER TEST LOG

LOCATION _____ DATE_____

TESTED BY _____

TANKER TEST RECORD					
TANKER	START	BEGIN FLOW	END FLOW	FINISH	LINES USED

NOTES:

Figure 5-21. Log sheet that can be used for recording the results of flow tests.

With a four-minute travel time, the rating is based on a distance of one mile from the water; this gives a basis for comparison with other tankers. Continuing with our example:

$$\frac{1{,}428}{6.75 + 3.75 + 4} = 98.5 \text{ GPM}$$

where

V = 1,428;
A = 6.75 minutes;
B = 3.75 minutes; and
T = 2 minutes

Use this test method only when the tanker has a pump for unloading. Tankers that have large dump outlets, either gravity or jet dumps, have to be tested using scales and a series of weight measurements.

WEIGHT TESTS

<u>Weigh a tanker to determine its efficiency.</u>
To find out whether a tanker can handle its load, weigh it using a truck scale. This way you can get a more accurate calculation of usable water, fill rate, and dump rate than flow tests provide. To get the data needed to make these calculations, you must weigh each tanker at least four times. Figure 5-22 is a typical worksheet for recording these

TANKER EVALUATION BY WEIGHT						
TANKER	INITIAL	AFTER DUMP	EMPTY	AFTER FILLING		
				FRONT	REAR	TOTAL

Figure 5-22. Worksheet that can be used for recording the weights when testing tankers.

APPARATUS EVALUATION WORKSHEET

APPARATUS EVALUATION BY WEIGHT

SPECIFICATIONS			
Fire Department	Grassy Pond	Unit No.	Tkr 31
Manufacturer	Chevrolet	Model/Yr	1991
WEIGHT RATING			
Gross Vehicle Weight Rating	33000	Axles	RA22000 FA11000
	TOTAL WEIGHT	FRONT AXLE	REAR AXLE	% WGHT FRT AXLE
W1 - On Arrival	30420			
W2 - After Max Dischge	15840			
W3 - Drained	15600			
W4 - After filling	30880	10080	20800	32.64
Percent of full load to GVWR (W4 or W1/GVWR X 100)				93.58
		WEIGHT	GALLONS	%CAPCTY
Full Load (W4 or W1 - W3)*		15,280	1,830	100.00
Spillage (W4 - W1)		460	55	3.01
Usable Water Initial (W1 - W2)		14,580	1,746	95.41
Usable Water Shuttle (W4 - W2)		15,040	1,801	98.42
Ballast (W2 - W3)		240	29	1.58

* Full load is to be determined by subtracting the weight when completely drained from the maximum weight recorded.

NOTES:

Figure 5-23. Worksheet for tanker evaluation by weight.

readings at the scales; Figures 5-23 and 5-24 can be used to record the pertinent information for each tanker as it is tested.

Make the first measurement when the tanker arrives at the site. Because the unit is considered to be in service, you can assume that the amount of water on board is the same amount it would have when it arrives at an incident scene. Record this value as Weight #1 (W1).

Then unload the tanker as rapidly as possible. If it has no dump outlet, use the pump to discharge the load at its maximum flow rate (see the section on making flow tests). If a dump outlet is the normal means of unloading the tanker, use it for the test. Open the dump valve at the

TEST RESULTS			
UNLOADING TESTS			
Pump Pressure		Dump Size	10" Square
Nozzle Pressure		Dump Type	Jet
Tip size		Dump Time	.55
Est./Msr Flow	399	Dump Rate	3,174.55
Discharge Time	4.38	Jet Pressure	
Usable Water I	1746	Usable Water I	1746
LOADING TESTS			
HYDRANT		DRAFT	
Supply	30	Supply	
Fill Lines	1 3"	Fill Lines	
Fill Pressure	100	Fill Pressure	
Fill Time	2.9	Fill Time	2.9
Fill Rate	631.03	Fill Rate	631.03
Capacity	1830	Capacity	1830
ESTIMATED CONTINUOUS FLOW CAPABILITY - USABLE WATER S			1801
DISTANCE	PUMPING	TRAVEL TIME	DUMPING
1 Mile	145.62	4.0	213.86
2 Miles	113.50	7.5	151.07
3 Miles	92.99	11.0	116.79
5 Miles	68.30	18.0	80.33

NOTES:

TESTED: 6-20-92 At Gaffney by William F. Eckman

Figure 5-24. Worksheet for calculating tanker capabilities from weight measurements.

beginning of the test, and leave it open until all of the usable water is discharged. At this point, close the dump valve and record the time of discharge. (Figure 5-21 is an example of a log for recording these times.) The exact time the valve should be closed is somewhat subjective and can be a source of inconsistency, but experience has shown that it is a reasonably well defined point. The reason that you close the valve before all of the water discharges is to maintain a high average flow rate. As the water leaves the tank, the level drops and head pressure decreases. Since the flow rate is a function of the size of the opening and the head pressure, the flow rate drops proportionately. It often takes as little as one minute to remove 90% of the water from a tank, but another full minute to discharge the remaining 10%. In most cases, it is detrimental to water shuttle operations to wait for that last 10%. Not only do you lose time, but the flow rate is too low to sustain the fire attack. In general there are two indications that it is the optimum time to terminate flow. One is when the flow through the dump outlet no longer fills the pipe; the other is the reach of the stream. Even with the variables, the discharge time recorded here still gives a more accurate indicator of expected dump time in water shuttle operations than the unloading time recorded while discharging through the pump.

After you discharge the usable water, weigh the tanker again. Record this figure as Weight #2 (W2). There will still be some water left in the tank. Open all drains, remove all caps, and let the water drain until the tank, pump, and piping are completely dry. Weigh the tanker again and record the empty weight as Weight #3 (W3).

When the tank is empty, fill it as rapidly as possible using the procedures detailed above under flow tests. Record fill time as determined from the time water begins to flow into the tank until it overflows. Weigh the tanker again and record the reading as Weight #4 (W4).

By using these four weight readings, you can make some accurate calculations of loading and unloading flow rates. It's also possible to analyze each tanker's strengths and shortcomings based on the amount of usable water carried versus ballast that serves no useful purpose. Often, relatively minor improvements increase the amount of usable water from a tanker appreciably once you analyze the test results in depth.

FULL LOAD

Calculate the total amount of water the tanker carries.

The full load measurement is the total amount of water a tanker can carry. Theoretically, it should be the difference between the full weight (W4) and the drained weight (W3), but not always. For one reason or another, some tankers weigh more when they arrive than they do after they are filled. When this happens, determine full load by taking the difference between the arrival weight (W1) and the drained weight (W3). As with all these measurements, convert the weight difference to gallons of water by dividing by 8.35, the weight of one gallon.

SPILLAGE

Determine the amount lost during travel.

Figure spillage by comparing arrival weight to the after-filling weight. Spillage generally represents the water lost through overflow as a tanker moves across the highway. How much is lost depends on the tank construction, primarily the design and location of the venting arrangement. The amount of spillage can really add up if a tanker responds to a number of alarms where no water is used and the tank is not refilled. Many tested tankers have shown inaccurate spillage figures because water was used from the tank for a small fire, a washdown, or other purpose, and not replaced before the unit went back into service.

Some tankers show a negative number for spillage, meaning that they have more water on board on arrival than they had after being filled. A negative number tends to be an indication of inadequate tank venting. If an air pocket gets trapped in the tank while water is coming in, the tank overflows before it is completely filled. The vents most likely are too small, but it is also possible that there simply are not enough openings in the top of the swash partitions inside the tank to allow air to move between compartments and exhaust through the vent. If you filled a tank slowly, for example, through a garden hose or a small line, it accepts more water because air can escape fast enough to accommodate the lower flow rate, whereas rapid filling (1,000 GPM means that 133 cubic feet of air had to be evacuated in one minute) overwhelms the venting system.

BALLAST

Analyze the amount of ballast a tanker carries.

When you closed the valve at the end of the rapid discharge test, some water was left in the tank. This water is referred to as ballast and it is called ballast because it serves no useful purpose. It is simply hauled back and forth in a shuttle, increasing the tanker's weight.

What conditions create ballast?

If you discharge by pumping, the whirlpool action at the tank outlet opening causes air to be drawn into the pump before the tank is empty, the same phenomenon that develops when you operate a pump from draft. As the water is drawn into the suction line rapidly, a swirling action can cause the pump to draw air prematurely. The pump goes into cavitation, leading you to believe that the tank is empty. And, if the tank to pump line lacks an anti-swirl device, this can happen while there is still an appreciable amount of water left in the tank.

Inadequate venting also can cause cavitation in the pump. As water leaves the tank, it has to be replaced by air, and if the air is not drawn into the tank rapidly enough through the vent, a partial vacuum or negative pressure can develop inside the tank. At the point of cavitation, the fire pump is operating with maximum vacuum at the eye of the impeller, and all of the pressure differential created by the vacuum in the pump, and atmospheric pressure in the tank, is used to overcome the friction loss of the water moving through the tank to pump line. If even a slight vacuum develops to subtract from the atmospheric pressure inside the tank, the pressure differential is reduced to the point that it is not able to overcome the friction loss in the line. Again, cavitation in the pump occurs while there is still water in the tank. Insufficient openings in the top of the swash partitions (baffles) also can cause this. If the vent and the tank to pump line are not in the same compartment, air probably cannot transfer fast enough; this causes a partial vacuum in that compartment, even if the vent is large enough to handle the flow rate adequately.

Swash partitions, also known as baffles, can create water transfer problems when you pump or dump. If water cannot transfer fast enough from one compartment to the other inside the tank, a whirlpool develops in the tank to pump compartment while the other sections of the tank still have plenty of water. Being able to transfer water between compartments is even more critical when you use large dump outlets from the tank. Large gravity dump outlets or jet dumps can produce flow rates as high as 3,000 GPM, but you need large openings in the baffles to supply these large flows through the dump valve. If the openings are too small, the dump will not flow a "full pipe" as the level in the compartment where the dump outlet is located drops, even though the level, and the resulting head pressure, in the other compartments, are still relatively high.

Determine the amount of ballast by subtracting the drained weight (W3) from the discharged weight (W2).

USABLE WATER

Determine the usable water in the tank.

Usable water is the amount left after the water that was lost in transit and the water trapped inside the tank are subtracted from the tank's capacity. This figure is much more significant than the tank's capacity because it is the amount of water you will have to use each time the tanker arrives at the fire scene. The amount of water available for your use when the tanker initially arrives on scene is different from the amount of usable water available on subsequent trips in a water shuttle.

Calculate the usable water on the initial trip by subtracting the weight after discharge (W2) from the weight on arrival (W1). Once the shuttle is operational, determine usable water by comparing the weight after filling (W4) with the weight after discharge (W2). Unless the weight after filling happens to be the same as the on-arrival rate, these two figures will be different. In extreme cases, the difference can be as much as 10%. The usable water in a shuttle is a more significant figure than the initial measurement since this is what the tanker will be able to deliver on a sustained basis as the shuttle operation progresses. The usable water in a shuttle would compare to the hydraulic capabilities of a water main when calculating the available fire flow for a particular location.

ANALYSIS

Identify deficiencies and develop a plan for improvement.

Analyzing weight test results not only is helpful in assessing a tanker's efficiency, it also offers information about how to improve it.

Comparing usable water, spillage, and ballast to total capacity expressed in percentage of full load is helpful. For comparison purposes, use the ISO calculation of 90% of the total capacity as usable. If your tanker can carry more than 90% usable water, it exceeds the ISO's expectations. If it cannot, it needs some improvement. Reasonably well-designed and constructed tankers can easily exceed 95% usable water. The tanker shown in Figure 5-16 had a 98.42% usable rate in a shuttle when it was tested, and 95.41% of a full load when it arrived at the test site. If this is not the case with your unit, refer to the spillage and ballast percentages for ideas on improvements.

OPERATIONAL TESTS

While we have rated mobile water supply apparatus assuming a handling time at each end of the shuttle of less than a minute to connect and maneuver, you can get a more accurate estimate by actually going through the evolution.

One way to do this is to follow the procedures the ISO uses when it evaluates a fire department's supply capabilities. To test unloading capability, set up a portable tank at a simulated fire scene. Place the tanker in position 200 feet away from the tank. On a signal, move the tanker into position, dump its load, and return it to the starting line. Record the time for this evolution as the unloading time. You can use the worksheet shown in Figure 5-21 to record the test results. Measure loading time in a similar manner. Move the supply pumper into position with fill lines laid, the pump primed, and the tanker stationed 200 feet away from the filling station. On a signal, move the tanker into position, connect the fill lines, fill the tank, and return it to the starting position. Use the total elapsed time as your actual loading time for each trip. If you differentiate between the time water was flowing and the time for getting into position and making connections on these worksheets, you can analyze the results of tests and see where improvements are needed. The loading and unloading times from the operational tests, the usable water from the weight measurements, and the travel time based on an average possible speed, give a very accurate measure of what each tanker's expected continuous flow capabilities would be when it operates in a water shuttle to supply a given location. This information gives a benchmark for determining whether fire protection throughout a fire management area that depends on tankers for water supply is adequate.

6. Fire Hydrants As A Water Source

OVERVIEW

A FIRE HYDRANT, located on an approved public water system, is generally the most reliable source of water for fire protection. Some private water systems also include fire hydrants, but they may not have flows sufficient for the structures they are installed to protect.

The ability of any water system to provide the needed flow for an adequate level of fire protection is limited by four factors:

- Maximum production rate of the wells or other water sources;
- Amount of water in storage;
- Distribution system; and
- Hydrant installation and maintenance.

PRESSURIZED SYSTEMS

The most common form of pressurized system, including fire hydrants, is a public water system, which a government agency generally is given the responsibility to operate. In some cases, a private company or quasi-governmental authority is responsible for the day-to-day operations of the publicly owned systems. In either case, the water company's primary objective as a public utility is to deliver potable water to its customers. It has to cover its expenses by imposing user charges and, if possible, try to make a profit. Fire protection is the system's secondary purpose. Installing hydrants and maintaining them tend to be low priorities; hydrants do not pay a water bill! Even where the local government pays a fire hydrant rental fee to the water company, the people in charge of the day-to-day operations often know little about the fire service's needs and do not place enough importance on fire hydrant maintenance. If a public water system is to perform adequately in terms of fire protection, fire department officials, particularly water supply officers, must maintain a good working relationship with the people who operate the water system.

Private systems sometimes are provided when an industrial, commercial, or institutional installation creates a serious fire and/or life hazard in an area where no public water system is available. These systems can range from small commercial complexes with a single storage tank, one yard hydrant, and an assortment of standpipe hose stations, to extensive installations that are larger than many public water systems. Some fire departments have operating procedures that do not allow pumpers to connect to yard hydrant systems as a source of water for firefighting. There are a number of reasons for not relying on private water systems.

First, the distribution system often uses pipe that is too small to supply the needed fire flow.

Second, the distribution system frequently is installed in a straight line configuration or a loop with no cross connections to provide a dependable flow to the hydrants from more than one direction.

Third, storage capacity generally is limited, with poor static pressure; this creates a low residual pressure at the hydrant when water begins to flow.

Finally, other demands on the system, e.g., built-in sprinklers, foam systems, etc., can interrupt or limit the flow to the attack pumper. Conversely, the attack pumper can deprive built-in systems of the water they need to function effectively.

Insurance companies often both initiate the installation of private water systems and maintain a degree of supervision over their operation. At the same time, the fire department also needs to familiarize itself with the system's capabilities and monitor it continually to make sure that it is ready and available when needed. Private systems require maintenance equal to that of public water systems.

A good water system is able to produce and store enough water to meet the total water supply needs of the area it serves, and its distribution system provides an adequate rate of flow throughout the area to be protected.

TOTAL WATER SUPPLY

The total water supply available for fire protection in any public water system is limited by the system's normal usage. The maximum amount of water maintained in reserve for fire protection is the average daily minimum storage. When peak usage for normal domestic supply exceeds the maximum production rate, the average minimum storage will be lower than the system's total capacity. This is because the water level in the storage tanks will be lowered by the demand on the system at certain times.

Some systems are designed to maintain a certain amount of water in reserve for fire protection as shown in Figure 6-1. The storage tank for this purpose has two outlets. One outlet feeds the domestic system and is located higher in the tank; a portion of the water is not accessible from this outlet. The hydrant distribution system is connected to the second outlet at the bottom of the tank, and all of the water is usable for fire protection. The position of the outlet to the domestic system determines what percentage of the water is available for normal usage, and how much is held in reserve until a fire hydrant is opened.

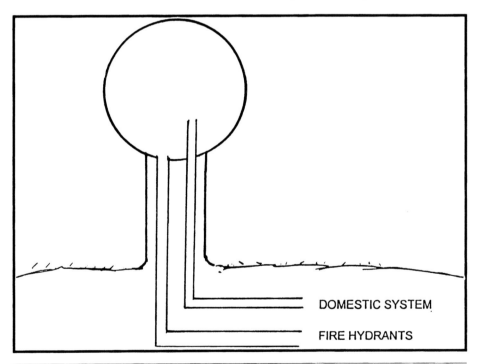

Figure 6-1. Elevated tank in a water system with a portion of its capacity reserved for fire protection.

STORAGE FACILITIES

Enough water must be stored in a central water system to equalize the flow and generally maintain an acceptable pressure during periods of peak demand from the system. In addition to the peak domestic usage, there should be enough water in storage to meet the expected demands for fire protection. This requirement can be met either by high-pressure or low-pressure storage facilities.

High-Pressure Storage

High-pressure storage facilities provide and maintain the necessary operating pressure throughout the distribution system. Such a system should be designed so that the pressure at the highest plumbing fixture in multistory buildings is at least 20 PSI when the level in the system is at its lowest operating point. The static pressure in the system should range from 50 to 75 PSI, and the lowest residual pressure in the distribution network should be at least 30 PSI. The average pressure in a well-designed system should be approximately 45 PSI during periods of peak consumption. To operate within these limitations, high-rise buildings more than three stories high may have fire pumps or internal storage systems to meet the pressure requirements on upper floors. From a fire protection standpoint, the available flow from a fire hydrant is calculated with a residual pressure of 20 PSI. Once the residual pressure drops below 20 PSI, no additional water should be drawn for fire protection. The system should be designed to supply the needed fire flow without reducing the pressure below this point.

High-Pressure Systems

High-pressure storage systems maintain pressure in one of the three ways illustrated in Figure 6-2.

Storage tanks within the distribution system are elevated aboveground on a supporting structure as shown in Figure 6-3. The minimum pressure in the system is determined by the distance from the bottom of the tank to the ground, and the ground level elevation compared with the rest of the system.

The second means is reservoirs, low-level storage tanks, or standpipes located at a higher elevation than the distribution system. Pressure in the system is dependent on the difference in ground level elevation at the storage site and the distribution system. Figure 6-4

Fire Hydrants as a Water Source 129

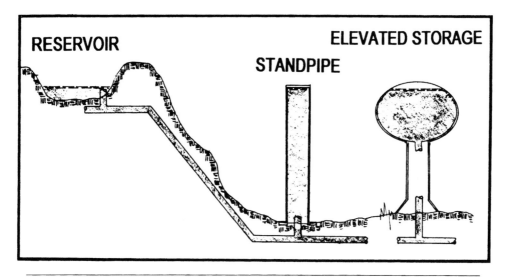

Figure 6-2. Types of storage in a public water system.

Figure 6-3. Elevated storage tank.

Figure 6-4. Standpipe storage tank.

Figure 6-5. Well house and Hydro-Pneumatic pressurized storage tank.

depicts a standpipe type tank. Systems in mountainous terrain may have to use pressure-reducing valves to prevent excessive pressure to the portion of the system at the lowest elevation.

With hydro-pneumatic or pressure tanks, pressure in the system is developed by high service pumps that force water into a pressurized tank. One of these tanks is shown in Figure 6-5. The pressure is maintained within certain limits by the compressed air inside the tank and the pumps.

Storage in the form of elevated tanks within the distribution system that are readily available and do not require mechanical or electrical equipment is advisable for fire protection purposes. The static pressure at any particular hydrant is determined by the difference in elevation between the surface of the water in the tank and the outlet on the hydrant. Figure 6-6 shows how the pressure in a system supplied by an elevated tank will vary from hydrant to hydrant. In this example, Hydrant A has a higher static pressure than Hydrant B.

Reservoirs or ground level storage can be substituted for elevated tanks where the terrain is suitable for this purpose. The difference in elevation between the point of ground level storage and the rest of the system must be great enough to generate the natural head pressure necessary to maintain a good working pressure in the distribution system as the level in the storage facility changes with the water usage. The three different means of attaining needed head pressure are shown in Figure 6-2. In this case, the reservoir's elevation is high enough to provide the required head pressure in the system.

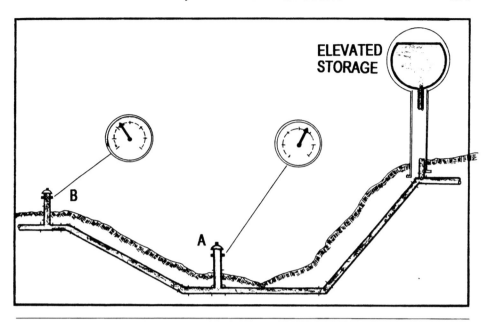

Figure 6-6. Pressure varies in a water system wih changes in elevation.

When standpipes, ground level storage tanks, or reservoirs are used in locations where there is not enough difference in ground level elevation, the pressure in the system will drop drastically as the water is used and the level in the tank is lowered (see Low-Pressure Storage, below). In some systems, 50% or less of the water in storage can be used for fire protection before the residual pressure at the hydrant drops below 20 PSI. When pressurized storage tanks are used, the pressure drop as the level in the tank is lowered means that as little as 18% of the total capacity of the tank can be used for fire protection purposes. In either case, the tank must be designed to keep enough usable water in reserve to meet fire protection requirements.

Low-Pressure Storage

Low-pressure storage uses low-level storage tanks or reservoirs for storing water; high service or fire pumps provide the needed pressure. Water is stored at or near atmospheric or gauge zero pressure, as illustrated in Figure 6-7, but is readily available for use. A low-pressure system can be used for fire protection, with the fire pumps engaged only when activated by an alarm system, but cannot be adapted to provide domestic water. In a public water system, low-pressure storage may be used to supplement the high-pressure storage system. During periods of peak demand, high-capacity pumps are used to boost the pressure

Figure 6-7. Ground-level storage tank.

to force the water into the high-pressure system and increase the available supply. If a fire protection system depends on low-pressure storage to meet the needed fire flow, some type of emergency power must be available to ensure the system's dependability.

Where a ground level or low-pressure storage tank is used, hydrant flow rates are limited by the pump's ability to deliver water fast enough to maintain the system's working pressure and by the size of the distribution system. In some cases, separate fire pumps, that start automatically when pressure drops, are provided for emergency situations.

SUPPLY WORKS CAPACITY

Estimates of total supply works capacity for ISO evaluations should include the minimum daily average storage, supply pump capacity, emergency supplies, suction sources, and any supplemental supply that the fire department can maintain. The 1980 ISO Rating Guide specifies that the supply works capacity of a system should be sufficient to maintain fire flows of up to 2,500 GPM for 2 hours, up to 3,500 GPM for 3 hours, and greater flows for periods based on the degree of risk.

DISTRIBUTION SYSTEM

There is a natural inclination for fire department personnel to assume that where hydrants have been provided, the water needed for fire protection will be available. **This is not necessarily true.** Many public water systems are very old, and cannot supply the amount of water needed for modern fire apparatus. When the systems were designed and installed, they were to protect buildings that were much smaller and required lower needed fire flows than today's buildings. Domestic consumption from the system also has increased significantly. Exacerbating the problem is that aging supply lines have deteriorated, significantly reducing the available flow from many older systems.

Some metropolitan areas have simply outgrown the capabilities of their water system. Even systems that were adequate when constructed have deteriorated over time with corresponding flow rate reductions. When supply lines are added to serve newly developed areas, water systems may not be able to furnish sufficient water.

A water system's static pressure is NOT a good measure of its capabilities. Static pressure indicates only the difference in elevation between the highest point in the system and the point of measurement

Figure 6-8. Small, outmoded storage tank.

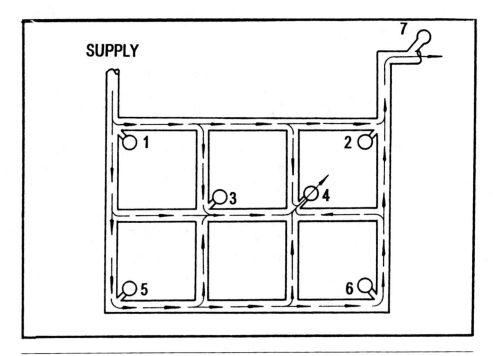

Figure 6-9. A water distribution system provides the most dependable supply for fire protection when it is "gridded" with frequent interconnections.

or the pumping pressure. Pressure developed by elevated storage tanks or pressurized tanks yields an average static pressure in the system of between 50 and 75 PSI. In mountainous terrain, static pressure in the system can be as high as 200 PSI because the reservoirs are at a higher elevation. At the same time, depending on reservoirs for pressure can result in a lower flow capability. The lower flow can be caused by the distance involved, or by friction loss in pipes that are too small or are in poor condition.

SYSTEM DESIGN

Friction loss in a system's water lines is primarily what limits water at various points. While pipe size is the single most important factor in designing a good water system, the configuration of the main distribution lines also contributes significantly to the end result. Figure 6-9 is a schematic diagram of a "gridded" distribution system, with hydrants installed throughout the network. The diagram shows the movement of water through the system with either of two hydrants, number 4 and number 7, opened and flowing water.

Hydrant number 7 is located on a "dead-end" main, and, as Figure 6-9 shows, all of the water taken from this hydrant has to come through the single line that supplies it. This hydrant's maximum flow is limited by the size, length, and condition of the main. Not only do dead-end mains tend to be installed with smaller pipe, but the flow through them generally is minimal. Dead-end mains tend to lose capacity when increased friction loss caused by accumulated sediment and corrosion inside the line takes place because of the limited flushing action of the low flow. Note that the flow from this hydrant could be reduced further if a second pumper connected to the system took water from hydrant number 2. Additional flow at this point would reduce the residual pressure in the system and restrict the amount of water available at hydrant number 7 as well.

Hydrant number 4, on the other hand, could be expected to have a much higher available flow rate from the system. Located within the grid, water can flow to this hydrant from a number of different directions, as shown by the arrows. The mains within the primary distribution grid generally are larger than the peripheral mains. In a gridded system, the flow is divided among several lines, so the peak flow in any one line and the accompanying pressure loss from friction within the pipes will be much less. In a gridded system, it usually is possible for more than one pumper to take water at the same time.

When a system includes dead-end mains, connecting them back into the primary grid is an improvement. In the system depicted in Figure 6-9, it might be cost effective to connect from the end of the line at hydrant 7 back into the system between hydrants 2 and 6. This would both provide increased flow by allowing hydrant 7 to draw from two different directions, and also ensure better water movement in the system while minimizing deterioration from sedimentation and rust or corrosion in the lines. From a domestic usage standpoint, the added movement of water in the looped system would significantly improve the quality of water to the users.

Larger dead-end water lines frequently are installed to meet fire protection requirements in commercial or industrial installations at isolated locations where little water is used during normal operations. With low domestic usage, it is difficult to maintain enough chlorine and other additives to make water safe for human consumption. To maintain water of acceptable quality, lines must be flushed periodically. Closing the loop eliminates the need to waste large amounts of water in flushing the line, and the cost savings could pay for additional pipe over a period of time.

When a system has only one storage tank, additional fire flow can be expected when the supply pumps, frequently connected to another point in the system, are running. Consider making arrangements with the water system's operator to have all pumps turned on manually when a major fire occurs and water demand is likely to be high.

You must know the design and capabilities of your water system to perform prefire planning, and to maintain adequate fire streams during major fires. The only way to verify available fire flow throughout the system is to conduct regular flow tests on hydrants as recommended in NFPA 291, *Recommended Practice for Fire Flow Testing and Marking of Hydrants*. It is the fire department water supply officer's responsibility to see that this is done.

HYDRANT INSTALLATION AND CONSTRUCTION

Hydrants should be installed at strategic intervals throughout all public water systems, but because they are expensive, the number of hydrants frequently is reduced during construction with the idea that they can be added later when more development takes place. All too often this never happens.

A common practice is to install fire hydrants at 1,000-foot intervals. To obtain an ISO Public Protection Classification (PPC) rating better than Class 9, the miminum requirement is that structures must be within 1,000 feet of a hydrant. Having hydrants located every 1,000 feet easily will meet the minimum requirements. However, keep in mind two often overlooked ISO stipulations.

First, the distance to the hydrant is measured as "hose can be laid," not in a straight line from the structure.

Second, the maximum credit for flow from any single hydrant is

- 1,000 GPM when the hydrant is within 300 feet of the location being evaluated;
- 670 GPM when the hydrant is within 301 to 600 feet of the location being evaluated; and
- 250 GPM when the hydrant is within 601 to 1,000 feet of the location being evaluated.

Since the maximum credit allowed for any one hydrant is 1,000 GPM, and the hydrant must be within 300 feet of the building being evaluated to receive full credit, it is clear that hydrants should be located at much closer intervals in commercial and industrial areas than in single-family residential neighborhoods. Such large buildings

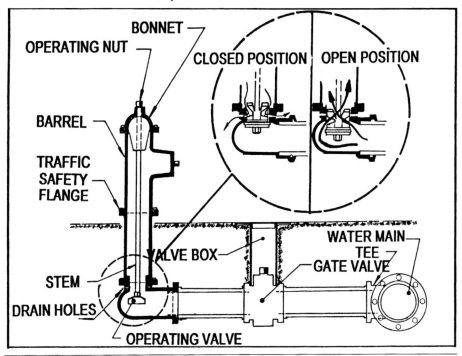

Figure 6-10. Diagram of standard fire hydrant showing the component parts.

as hospitals, nursing homes, and schools that present a serious life hazard should have more than one hydrant installed within 300 feet of each building at locations readily accessible to fire department apparatus. Even in residential neighborhoods, hydrants must be located so that no structure will be more than 600 feet from the closest hydrant or within 1,000 feet of two hydrants. This is the case if the 500-GPM fire flow that is the minimum requirement set by the ISO for single-family residences is to be credited.

HYDRANT OPERATION

Where freezing temperatures are common, all hydrants approved for installation by the Underwriters Laboratories (UL) for use in these areas are equipped with a valve that drains the barrel of the hydrant each time that it is closed. Figure 6-10 is of a typical hydrant and shows the location of the drain valve. The operating shaft of a fire hydrant is designed so that when the main valve to the system is closed, the drain valve on the hydrant is open. When the main valve is open, as shown in Figure 6-10, the drain valve is closed. This happens only when the hydrant is fully opened or fully closed. If the hydrant is not fully

Figure 6-11. Above, left, cutaway view of standard fire hydrant showing the drain and method of installation.

Figure 6-12. Right, hydrant in the process of installation showing the valve, the victaulic coupling, and the thrust rods.

opened during use, it is possible for the drain valve to be open while the main valve from the system is open. When the drain holes are open, water escapes under pressure and causes the ground to wash away from the base of the hydrant. Over time, hydrants can break because of the lack of support, and become unusable. On the other hand, if the hydrant is not fully closed, the drain valve will not open and allow the water to drain from the barrel of the hydrant. In warm weather, this is not problematic; however, if the hydrant is not fully drained during winter months, the water in the barrel can freeze, rendering the hydrant unusable in an emergency.

Water hammer, caused when water flow is suddenly cut off, is damaging to the system and a potential problem for hydrant users. Standard hydrants are designed to require a minimum of ten full turns on their operating shafts to open or close their main valves. This design feature prevents changes in flow from occurring so rapidly that it damages the system. Any time a pumper or an external valve that can be opened or closed rapidly, such as a quarter-turn ball valve or four-way hydrant valve, is connected to a hydrant, there is potential for water hammer to occur. Figure 6-12, a photograph of a hydrant under installation, shows the thrust rods which help to absorb force. Regardless, water hammer can still damage a hydrant beyond use. Opening or closing a valve too quickly may cause serious damage to the system, and extreme caution should be used.

HYDRANT TESTING

Improper hydrant testing can damage water systems and private property, cause injuries and create fire department liability and public relations problems. Try to avoid potential problems by using the following precautions.

Hydrant Testing Safety Precautions

- Never conduct flow tests without the authorization of the water system supervisor. If possible, have someone from the water system's maintenance department present to operate the hydrants while the tests are being conducted.
- Minimize damage by directing the stream of water that flows from the hydrant (Flooding homeowners' lawns, shrubs, private property, or basements is both a liability and a public relations problem).
- Never direct a stream of water across a busy street without traffic control (Obtain police department help if possible).
- **Never discharge a stream of water into a street or roadway during freezing weather.**
- Check for water discoloration and continue to flush the mains until the water runs clear. Have water authorities warn area residents about water discoloration during hydrant testing.
- Never connect a rigid diverter directly to a hydrant outlet. A rigid diverter consists of a piece of pipe that extends to a desired length, then bends at an angle of up to 90 degrees to divert water flow. The reaction caused by the velocity of the water leaving the opening is considerable and can damage the hydrant. If the hydrant location requires that water be diverted, use either a diffuser or a section of flexible hose properly restrained at the point of discharge.

Figure 6-13. Diffuser being used to divert the stream away from the grass while flow testing a hydrant.

Figure 6-14. Set of hydrant test gauges.

Figure 6-15. Using a flow test kit for hydrant testing.

Hydrant Testing Equipment

1. Pitot gauge or flow meter.
2. Hydrant cap with pressure gauge.
3. Measuring scale calibrated to $\frac{1}{16}''$-markings.
4. Two hydrant wrenches.
5. Smooth bore nozzles in assorted sizes.
6. Flow test kit with various size nozzles and Pitot tube and gauge (optional, for convenience).

PRINCIPLES OF HYDRANT TESTING

To determine how much water is available from a hydrant system, compare the residual pressure when a measured amount of water is flowing to the static pressure in the system before the hydrant is opened. By determining the pressure drop with a known flow, the flow that can be expected for any given drop can be calculated. Flows generally are calculated for the pressure drop that allows a residual pressure of 20 psi.

Static pressure, as it is used in hydrant testing, actually is not static pressure. There is no way to cut off domestic consumption in a system during testing, so the static pressure, as measured, refers to a lack of fire flow; normal usage of water in the system is assumed. Given that water usage varies at different times of the day, this factor needs to be considered in any flow testing program. To minimize public reaction to possible water discoloration, some water systems perform flow tests between the hours of midnight and 6 a.m. However, the results of flow tests taken during this period often are misleading because the available fire flow during periods of peak consumption is significantly less than the amount available during periods of low usage.

These calculations are based on the fact that as water flows from a hydrant, friction loss in the system increases, reducing the residual pressure reading at the hydrant. To protect the system, a good practice is to keep the residual pressure at the hydrant at 20 PSI or higher; the maximum flow the ISO will credit must be attained with at least 20 PSI of residual pressure. This provides a margin of safety and helps to ensure that the pressure in the system never goes below 0 PSI, which would create a vacuum on the system. There are two reasons why the residual pressure in the system should not drop below 0 PSI.

The first, which you will recall, is damaging water hammer. The second is that if a negative pressure develops within the distribution system, contaminants could be forced into the system by atmospheric pressure, causing public health problems.

Gauges on a typical fire department pumper tend to be imprecise near the 0 PSI reading and, with all-too-common calibration problems, it's smart to maintain a reasonable margin of safety when you are operating from a hydrant. If you maintain a minimum residual pressure of 20 PSI, you'll have a margin of safety, and avoid creating a vacuum on the system.

The rating of a specific hydrant, or a particular point in a system, is based on the maximum flow achievable without a drop in residual pressure at that point below 20 PSI. One exception to this rule is a low-pressure hydrant system. If a hydrant's static pressure is less than 40 PSI, its flow rating is calculated for a residual pressure that is 50% of the static pressure instead of 20 PSI. To determine the maximum safe flow from a hydrant, pressure drop with a known flow is measured. Using these measurements you can estimate how much flow can safely be removed from the system at that point without reducing the residual pressure below 20 PSI or half of the static pressure.

TEST HYDRANT SELECTION

To test a system's total flow capabilities, a minimum of two hydrants are used (as shown in Figure 6-16). The hydrant closest to the desired test location becomes the test hydrant. Static and residual pressures then are measured by putting a tapped cap and gauge on one of the 2½"-outlets. You open the next hydrant on the line, in the direction of flow from the test hydrant, to establish the desired flow from the system. It is necessary to open more than one flow hydrant if the system supplies more water than one hydrant can discharge. You will realize that additional flow hydrants are needed when opening one hydrant does not cause a drop at the test hydrant of at least 10 PSI. Additional flow hydrants should be in the line of flow as depicted in Figure 6-16.

Normally, flow in the system is from larger lines to smaller lines, and from storage or production facilities toward the point of discharge. When you have a gridded system, several storage facilities, different sizes of water mains or any combination of these, it is more difficult to determine the direction of flow. When water can flow in several different directions, flow direction is not critical and the test still will be valid. If you need additional flow hydrants, the test hydrant should be located somewhere in the middle of the flow hydrants.

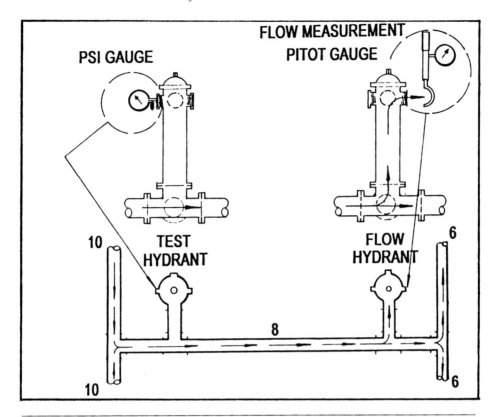

Figure 6-16. The flow hydrant should be beyond the test hydrant in the direction of flow in the system.

TEST PROCEDURES

Accurate measurements are best made by recording pressure readings at both hydrants simultaneously, while communicating by radio between operators at the flow and test hydrants.

How to Test a Hydrant

1. Remove one of the 2½"-caps from the test hydrant and install the tapped cap and gauge.
2. Open the test hydrant fully and allow the air to bleed off from the bleeder valve on the test gauge.
3. After the pressure stabilizes, record the static pressure on a worksheet (such as the one included in Figure 6-18). Be patient; it often takes some time before the pressure stabilizes enough to take an accurate reading.

Figure 6-17. A cap gauge can be used to measure the static pressure at the test hydrant before any water is flowing.

4. Remove one of the 2½″-caps from the flow hydrant. Measure the diameter of the opening to the closest ¹⁄₁₆″. Feel the inside of the opening to determine the coefficient of the hydrant as demonstrated in Figure 6-19. Record the hydrant coefficient for use in making flow calculations.
5. Open the flow hydrant fully. Contact the person at the test hydrant so readings can be coordinated. Take a reading with the Pitot gauge by inserting the tip of the tube into the center of the stream a distance equal to one-half of the diameter from the opening as depicted in Figures 6-20 and 6-21. Record the Pitot reading.
6. Record the residual pressure on the gauge at the test hydrant while the reading is taken of the discharge from the flow hydrant. If the residual pressure has not dropped at least 10 PSI, additional flow will be required to make a large enough change to provide an accurate estimate of what the flow would be at 20 PSI residual pressure in the system. Close the flow hydrant and remove the cap on the second 2½″-outlet, and measure the outlet size. Open the hydrant again and allow water to discharge from both outlets.

Fire Hydrants as a Water Source

HYDRANT TEST RECORD

TEST LOCATION_____

HYDRANT TEST DATA							
HYDRANT NUMBER	STATIC PRESS.	RESID. PRESS.	PITOT PRESS.	FLOW OPENING	FLOW TABLE	HYDRANT COEFF.	GPM FLOW

TEST LAYOUT

CALCULATIONS

$$GPM \times \frac{PDC}{PDT} = FLOW$$

PD	K	PD	K	PD	K	PD	K	PD	K
1	1.00	21	5.18	41	7.43	61	9.21	81	10.73
2	1.45	22	5.31	42	7.53	62	9.29	82	10.80
3	1.81	23	5.44	43	7.62	63	9.37	83	10.87
4	2.11	24	5.56	44	7.72	64	9.45	84	10.94
5	2.39	25	5.69	45	7.81	65	9.53	85	11.01
6	2.63	26	5.81	46	7.91	66	9.61	86	11.08
7	2.86	27	5.93	47	8.00	67	9.69	87	11.15
8	3.07	28	6.05	48	8.09	68	9.76	88	11.22
9	3.28	29	6.16	49	8.18	69	9.84	89	11.29
10	3.47	30	6.28	50	8.27	70	9.92	90	11.36
11	3.65	31	6.39	51	8.36	71	9.99	91	11.43
12	3.83	32	6.50	52	8.44	72	10.07	92	11.49
13	4.00	33	6.61	53	8.53	73	10.14	93	11.56
14	4.16	34	6.71	54	8.64	74	10.22	94	11.63
15	4.32	35	6.82	55	8.71	75	10.29	95	11.69
16	4.48	36	6.93	56	8.79	76	10.37	96	11.76
17	4.62	37	7.03	57	8.88	77	10.44	97	11.83
18	4.76	38	7.13	58	8.96	78	10.51	98	11.89
19	4.90	39	7.23	59	9.04	79	10.59	99	11.96
20	5.04	40	7.33	60	9.12	80	10.66	100	12.02

Tested by_____ Date_____

Figure 6-18. Static pressure should be recorded on a hydrant test record worksheet.

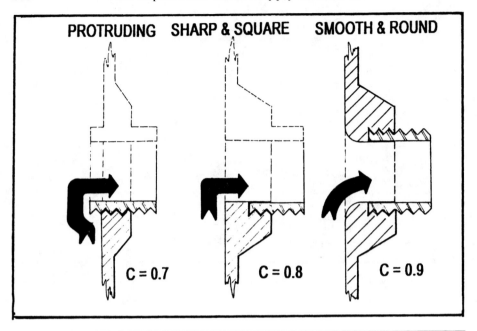

Figure 6-19. The coefficient of flow from the 2½"-outlet of a fire hydrant varies with the type of construction.

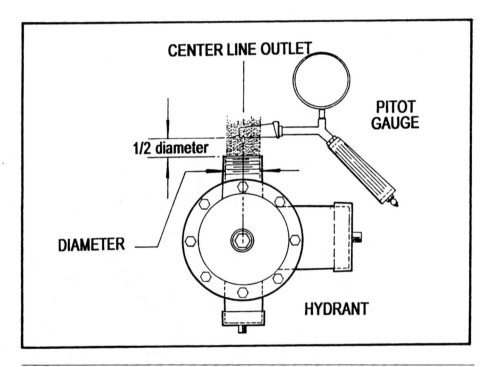

Figure 6-20. To measure the flow from a hydrant, the Pitot Blade should be in the middle of the stream, a distance of 1/2 the diameter from the outlet.

Figure 6-21. *A hand-held Pitot gauge can be used to determine the discharge from the flow hydrant while a residual pressure reading is recorded at the test hydrant.*

Figure 6-22. *A hydrant test gauge provides a more accurate reading of the flow from a hydrant.*

7. Take a reading with the Pitot gauge on both of the openings of the flow hydrant. Record both readings in order to determine the total flow.
8. Record the residual pressure at the test hydrant again while measuring the flow from both openings on the flow hydrant. If the residual pressure still has not dropped at least 10 PSI, close the 2½″-openings on the flow hydrant and open the 4½″-opening, and repeat the test. Because there are irregularities in the flow from the 4½″-opening, flow readings will not be as accurate as they would be from the 2½″-outlets. If this large outlet is used, a correction factor, such as the one found in Table 1-11 of NFPA 291, can be applied to the results to make them more accurate. This, however, is not a perfect solution. More effective would be to flow water from several hydrants to get a significant pressure drop if the system is especially strong in the area where the tests are being made. What is important is to know exactly how much water is flowing through the system at the time of the test, and how much the pressure drops at the test hydrant.
9. After all of the readings have been recorded, shut down the flow hydrant and close all of the openings. Observe the gauge on the test hydrant, making sure that the static pressure returns to its original reading.
10. Shut down the test hydrant. Verify that the barrels have drained completely on all the hydrants that have been used for these tests.

COMPUTING THE RESULTS

The first step in evaluating the results of the flow tests is determining the amount of water that was discharging in gallons per minute. The formula commonly used to make this calculation is

$$29.83 \times d^2 \times \sqrt{P} = GPM$$

where
- d = Inside diameter of the discharge outlet
- P = Pressure at point of discharge measured with a Pitot Gauge
- GPM = Estimated flow in gallons per minute

For example, if the inside diameter of the outlet is exactly 2½″-and the reading on the Pitot gauge is 40 PSI, the calculation would be

$$29.83 \times 2.5^2 \times \sqrt{40} = \text{GPM}$$

$$29.83 \times 2.5 \times 2.5 \times 6.33 = 1{,}180 \text{ GPM}$$

Table 6.1 is a compilation of the expected flow from outlets of different diameters based on the velocity of the water leaving the outlet as measured with a Pitot gauge. Large diameter outlets were not included in this table because it is difficult to make accurate measurements from them using a Pitot gauge. If you must flow more water than a single 2½″-outlet can move, you should open additional outlets. In extremely good systems it is more accurate to use multiple hydrants than to open a large outlet to provide the needed flow. Pressure measurements under 10 PSI or greater than 50 PSI were not included in Table 6.1. When the pressure is less than 10 PSI, it is difficult, if not impossible, to get an accurate reading with a Pitot gauge. At the other extreme, enough water should be allowed to discharge during a flow test to bring the residual pressure as close to 20 PSI as is practical; remember, it should never exceed 50 PSI from a single outlet. Table 6.1 presents the range of pressures that normally will be encountered during flow tests.

All of these calculations were based on a nearly frictionless discharge orifice, although, in actual practice, this rarely happens. The actual flow from a discharge device can be as much as 99% of the theoretical discharge if you are using a well-designed smooth bore nozzle, or a water flow test kit. It can be as little as 70% (or less) of the theoretical flow from a poorly designed hydrant. By way of compensation, a coefficient of discharge is used to correct the theoretical reading and allow you to make a more accurate estimate of the actual flow from the outlet.

When you are conducting a flow test, it is the hydrant's construction that determines the coefficient of discharge. Figure 6-19 shows you three types of hydrant outlets and the expected discharge coefficient for each type. To allow for the discharge coefficient (C), the formula for determining flow has to be modified

$$29.83 \times d^2 \times \sqrt{P} \times C = \text{GPM}$$

If, in the example on page 27, the discharge coefficient is assumed to be .9, the new calculation would be

$$29.83 \times 2.5^2 \times \sqrt{40} \times .9 = \text{GPM}$$

$$29.83 \times 2.5 \times 2.5 \times 6.33 \times .9 = 1{,}062 \text{ GPM}$$

Table 6.1. Approximate Flow from Circular Outlets

Velocity PSI	Flow in GPM/Outlet Diameter in Inches					
	2¼''	2⅜''	2⁷⁄₁₆''	2½''	2⁹⁄₁₆''	2⅝''
10	478	532	560	590	619	650
11	501	558	588	618	650	682
12	523	583	614	646	679	712
13	544	607	639	672	706	741
14	565	630	663	698	733	769
15	585	652	686	722	759	796
16	604	673	709	746	784	822
17	623	694	731	769	808	847
18	641	714	752	791	831	872
19	658	733	773	813	854	896
20	675	752	793	834	876	919
21	692	771	812	854	898	942
22	708	789	831	874	919	964
23	724	807	850	894	939	986
24	740	824	868	913	960	1007
25	755	841	886	932	979	1028
26	770	858	904	951	999	1048
27	785	874	921	969	1018	1068
28	799	890	938	987	1036	1088
29	813	906	954	1004	1055	1107
30	827	922	971	1021	1073	1126
31	841	937	987	1038	1091	1144
32	854	952	1003	1055	1108	1163
33	868	967	1018	1071	1125	1181
34	881	981	1033	1087	1142	1199
35	893	995	1049	1103	1159	1216
36	906	1010	1063	1119	1175	1233
37	919	1023	1078	1134	1191	1250
38	931	1037	1093	1149	1207	1276
39	943	1051	1107	1164	1223	1284
40	955	1064	1121	1179	1223	1239
41	967	1077	1135	1194	1254	1316
42	979	1090	1149	1208	1269	1332
43	990	1103	1162	1223	1284	1348
44	1002	1116	1176	1237	1299	1363
45	1013	1129	1189	1251	1314	1379
46	1024	1141	1202	1264	1328	1394
47	1035	1154	1215	1278	1343	1409
48	1046	1166	1228	1292	1357	1424
49	1057	1178	1241	1305	1371	1439
50	1068	1190	1253	1318	1385	1453

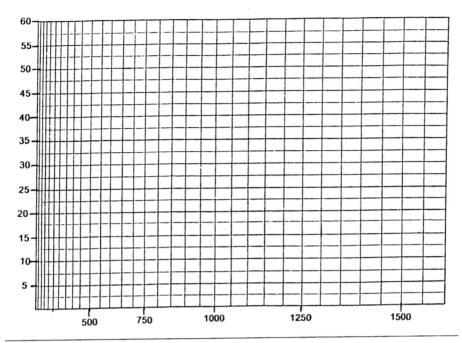

Figure 6-23. A flow worksheet can be used to determine the flow at different pressure points.

Although you can use the formula to calculate the exact flow, it is quicker and easier to use Table 6.1 or the types of charts and tables found in NFPA 291 to determine the flow from the hydrant, based on the Pitot gauge reading and the actual size of the discharge opening. The static and residual pressures then can be plotted on the vertical axis of the logarithmic graph paper on the "Hydrant Test Worksheet" shown in Figure 6-23, and the total flow or sum of flows plotted on the horizontal axis. In cases where more than one flow reading and residual pressure reading have been recorded, each of these points can be plotted on the graph. (It might be helpful to use separate graphs.) If all of the information is accurate, the line on the graph that connects the horizontal and vertical dots should be a straight line. Once you have drawn the line on the graph paper, you should be able to determine your flow at any given pressure. To display the test results, select the desired residual pressure on the vertical index line, and follow the corresponding horizontal line to the diagonal line that was drawn. Where the pressure line intersects the diagonal line, follow the vertical line at that point down to the flow reading. Use this method to determine the flow at any given residual pressure, or to determine the residual pressure for a given flow rate.

The Hazen-Williams formula is one mathematical method you can use to determine available fire flow; it has been incorporated into the hydrant test record worksheet, Figure 6-18. One variation of the Hazen-Williams formula that can be used to calculate available flow from a system is

$$\frac{\text{PD (Calculated)}}{\text{PD (Test)}} \times \text{Q (Test)} = \text{Q (Calculated)}$$

where

PD (Test)	=	Difference between the static pressure and the residual pressure during flow test
PD (Calculated)	=	Difference between the static pressure and the desired residual pressure
Q (Test)	=	Gallons per minute flowing during the test
Q (Calculated)	=	Gallons per minute flow capability at desired residual pressure

To use the pressure drop in this formula, the drop in PSI has to be taken mathematically to the 0.54 power. Table 6.2 lists a factor (F) that can be used to compute these calculations instead of using a complicated mathematical formula to carry the pressure drop figures to the 0.54 power. This formula allows the hydrant inspector to calculate the flow that can be expected at any residual pressure in the system more quickly and with greater accuracy than with a graph.

For example, assume that the static pressure was 45 PSI at a test hydrant before the flow test. With 707 GPM flowing, the residual pressure at the test hydrant dropped to 32 PSI. To determine what the maximum flow in the system would be at a residual pressure of 20 PSI, the formula was used to make the following calculations:

$$\frac{\text{PD (Calculated)}}{\text{PD (Test)}} \times \text{Q (Test)} = \text{Q (Calculated)}$$

$$\frac{5.69 \ (25 \text{ Drop})}{4.00 \ (13 \text{ Drop})} \times 707 \text{ GPM} = 1{,}005 \text{ GPM}$$

Using a graph to determine the flow at 20 PSI, as shown in Figure 6-24, yields the same results, but more precision is gained from using the formula versus the chart to make the calculations.

The procedure detailed here gives information about the maximum available fire flow a water system can support safely at a given location. It allows fire department personnel to take water from a number of

Fire Hydrants as a Water Source

Table 6.2. Pressure Drop Factors Taken to 0.54 Power

PD	F	PD	F	PD	F	PD	F	PD	F
1	1.00	21	5.18	41	7.43	61	9.21	81	10.73
2	1.45	22	5.31	42	7.53	62	9.29	82	10.80
3	1.81	23	5.44	43	7.62	63	9.37	83	10.87
4	2.11	24	5.56	44	7.72	64	9.45	84	10.94
5	2.39	25	5.69	45	7.81	65	9.53	85	11.01
6	2.63	26	5.81	46	7.90	66	9.61	86	11.08
7	2.86	27	5.93	47	8.00	67	9.68	87	11.15
8	3.07	28	6.05	48	8.09	68	9.76	88	11.22
9	3.28	29	6.16	49	8.18	69	9.84	89	11.29
10	3.47	30	6.28	50	8.27	70	9.92	90	11.36
11	3.65	31	6.39	51	8.36	71	9.99	91	11.43
12	3.83	32	6.50	52	8.45	72	10.07	92	11.49
13	4.00	33	6.61	53	8.53	73	10.14	93	11.56
14	4.16	34	6.71	54	8.62	74	10.22	94	11.63
15	4.32	35	6.82	55	8.71	75	10.29	95	11.69
16	4.47	36	6.92	56	8.79	76	10.37	96	11.76
17	4.62	37	7.03	57	8.88	77	10.44	97	11.83
18	4.76	38	7.13	58	8.96	78	10.51	98	11.89
19	4.90	39	7.23	59	9.04	79	10.59	99	11.96
20	5.04	40	7.33	60	9.12	80	10.66	100	12.02

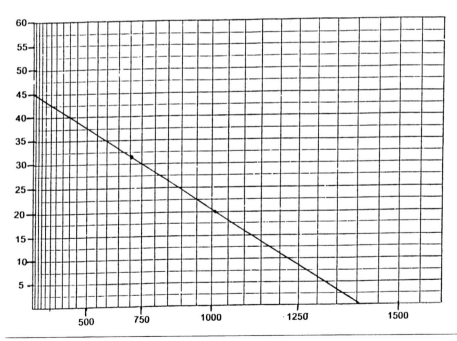

Figure 6-24. Flow worksheet developed for example on page 145.

hydrants accessible to the fire location without damaging the system, as well as indicating what the total available fire flow will be.

If you need to determine the flow from a specific hydrant, you can use the same hydrant for both test and flow measurements as shown in the photograph labeled Figure 6-25. Connect the tapped cap and gauge to one of the 2½''-outlets and follow the earlier procedures. Record the static pressure with the hydrant open and no water flowing. Close the hydrant and remove the cap from the other 2½''-outlet. Open the hydrant and take a reading with a Pitot gauge to determine the flow rate. Perform the calculations as you would in the two hydrant-method; the resulting flow rate will apply to the one hydrant only. If you do not have a Pitot gauge, read the cap gauge, assuming that the flow pressure from the discharge would be the same as the residual pressure on the hydrant, and look up the flow on the charts in NFPA 291 or on Table 6.1; this will give you a usable estimate of the actual flow from the hydrant.

Figure 6-25 offers another, more accurate way of determining flow. Because many hydrants deviate somewhat from the standard diagrams, it is difficult to determine many of their coefficients. One way around this problem is to use a flow test kit, as shown in Figure 6-25, with a stream shaper from a master stream device to reduce the turbulence

Figure 6-25. *Flow and test measurements can be taken at the same hydrant for an individual hydrant flow determination.*

and get a more accurate reading. With the flow test kit, the Pitot blade is always positioned accurately in the stream of discharge water, and is held firmly in position during the test. With the flow test kit come charts that give the measured flow when using a stream shaper and flow test kit. Take your reading directly from the chart, and the hydrant coefficient will not affect the total flow measurement.

USING A FLOW METER

It is simpler to perform hydrant flow tests with a flow meter instead of a Pitot gauge. An electronic flow meter, shown in Figure 6-26, can be used for this purpose. If a stream shaper or a 36″-length of pipe is used between the flow meter and the hydrant outlet connection, the results will be more accurate. The added pipe reduces the amount of turbulence as the water leaves the opening and provides a more stable and accurate reading.

HYDRANT COLOR CODING

After tests have been made throughout the water system, it is helpful to color code all hydrants to indicate each one's capacity. Each hydrant's barrel should be painted chrome yellow, with the hydrant tops and caps painted according to the standard color code in Table 6.3, established and recommended for use by the NFPA and the ISO.

Table 6.3. Standard Color Codes

Capacity	Color	Class
1,500 GPM or greater	Light Blue	AA
1,000 to 1,499 GPM	Green	A
500 to 999 GPM	Orange	B
Less than 500 GPM	Red	C

As part of your standard operating procedures, consider adding a rule that prohibits hooking more than one pumper to a hydrant in an area where the hydrants are color coded as orange or red. Systems of mostly green- or blue-coded hydrants should be able to support two or more pumpers operating from the system simultaneously.

Some state health departments have removed from public water systems hydrants that will not supply more than 500 GPM because of the

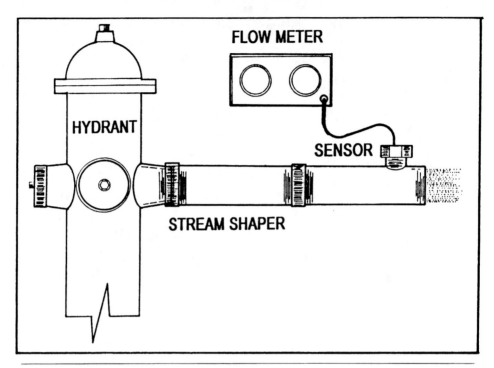

Figure 6-26. An electronic flow meter can be used instead of a Pitot gauge for testing hydrants.

Figure 6-27. Testing the flow from a fire hydrant with an electronic flow meter.

probability that a fire department pumper connected to a low-capacity hydrant could inadvertently create a vacuum in the system and contaminate the potable water supply. Fire departments should never connect hard suction hose to a fire hydrant of questionable capacity. Soft hose connected from the pumper to a hydrant will collapse and cause the pump to go into cavitation before causing a vacuum in the water system.

IN-SERVICE ESTIMATIONS OF CAPACITY

If you do not know a particular hydrant's flow rating, or if you have no way of knowing how much water already is being used at the time you connect, the pump operator can estimate the remaining capacity of the hydrant. To do so, compare the static pressure registered by the compound gauge at the intake of the pump when you first connect to the system with the residual pressure reading after water begins to flow.

If the residual pressure on the pumper's intake drops 10% below the static pressure you observed when you first connected to the hydrant, it means that one-third of the available water supply is being used. It also means that two additional lines of the same size flowing the same amount of water could be supplied. If the difference is 25%, one-half of the water is being used, and only one additional line could be supplied safely. If the difference is more than 25%, any additional flow would be limited to less than the amount being used.

The following is an example of how the pressure drop can be used to estimate the available water supply when operating from a hydrant.

A hydrant measured 80 PSI static pressure when the pumper was connected. A 2½"-attack line flowing 250 GPM was put into service. The residual pressure dropped to 72 PSI. To calculate the remaining available flow

$$\frac{\text{Static Pressure} - \text{Residual Pressure}}{\text{Static Pressure}} \times 100 = \text{Percent Drop}$$

$$\frac{80 \text{ (Static)} - 72 \text{ (Residual)}}{80 \text{ (Static)}} \times 100 = 10\% \text{ Drop}$$

A 10% change indicates that one-third of the water is being used and that two more lines could be supplied or an additional 500 GPM would be available.

If the pressure had dropped to 60 PSI when the initial attack line was put into service, the calculations would change

$$\frac{80 \text{ (Static)} - 60 \text{ (Residual)}}{80 \text{ (Static)}} \times 100 = 25\% \text{ Drop}$$

A 25% change means that one-half of the available water is being used, and only one additional line could be supplied for an additional 250 GPM.

As additional hoselines are put into service and other pumpers are connected to the system, the pump operator must note any changes in residual pressure; these changes indicate that the available supply is limited. It is possible for a water system to become overextended to the point that additional pumpers simply take water away from those already operating. An alert pump operator tells the water supply officer about potential problems in enough time to prevent interruptions to the existing water supply.

HYDRANT MAINTENANCE

Fire hydrants are used primarily to supply the water for fire department emergencies. Generally used under emergency conditions, all hydrants in a system must be properly maintained and ready for use at all times. Although it is the fire department that uses fire hydrants, it is the water utility's responsibility to maintain them. In some cases, water departments have entered into agreements that allow fire departments to do the required testing and maintenance. While this enables the fire department to ensure good hydrant maintenance, the responsibility for system maintenance rests ultimately with the owner. The American Waterworks Association's (AWWA) *Manual of Water Supply Practices* (AWWA M17) explains both the specifics of setting up a hydrant maintenance program and how to perform the required maintenance.

As part of its process of assigning a Public Protection Rating to a fire department, the ISO reviews hydrant inspection and maintenance records in the water supply portion of the evaluation. The ISO recommends inspecting each hydrant in a system every six months; nearly 10% of the credit for water supply depends on this inspection, based on the records that are available. An example of a hydrant test and maintenance worksheet is shown in Figure 6-28.

HYDRANT RECORD

HYDRANT NUMBER

LOCATION:

HYDRANT INFORMATION:

MANUFACTURER_____OUTLETS_____

DATE INSTALLED_____BY _____

ALTITUDE_____SIZE OF MAIN_____

DATE	TESTER	HYDRANT PRESSURE AND FLOW TEST RECORD					
		HYDRANT TEST (one hydrant used for both flow & test)			SYSTEM TEST (second hydrant used for flowing water)		
		Static Pressure	Flow Pressure	GPM Flow	Static Pressure	FLow Pressure	GPM Flow
		RECORD OF MAINTENANCE AND REPAIRS					
		WORK ITEM	DATE WORK PERFORMED			
		LUBRICATED					
		INSPECTED					
		TESTED					
		REPAIRED -LIST					

Figure 6-28. Hydrant test and inspection test record.

INSPECTION

All hydrants should be inspected regularly: the ISO recommends twice-yearly inspections, while the AWWA specifies a minimum of one inspection per year. In addition to routine inspections, fire departments should notify their water department after any extended use of a hydrant. This is an opportunity to reinspect for damage and check its operability.

In freezing climates, hydrants should be inspected in the fall and again in the spring. In the fall, check to make sure that the drain valves work on all hydrants, and that hydrant barrels have been drained completely after the inspection. In the spring, a physical inspection should reveal any damage from freezing and thawing of the ground or from water that was not drained completely after hydrant use.

Inspection Procedures

Your inspection begins with a physical examination of the hydrant and its surrounding area to identify any potential difficulties the fire department might have when connecting to, and using, the hydrant. This very thorough inspection should, at a minimum, include

- Checking and verifying the hydrant location to make sure it matches the information on the hydrant test and maintenance record worksheet.
- Checking for any obstructions that could block access to the hydrant, including signs, traffic barriers, utility poles, fences, shrubbery or vegetation, etc.
- Checking ground clearance to verify that there is sufficient unobstructed space between the bottom of the pumper outlet and the ground to allow the fire department to connect its suction hose without having to use a shovel to dig it out. Figure 6-29 is a photograph of a hydrant located on a street that was overlaid until the pavement covered the outlets.

Figure 6-29. Hydrant buried in the street after paving has been completed.

Fire Hydrants as a Water Source   161

Figure 6-30. Accumulation of dirt around hydrant would make it difficult to use.

Figure 6-31. Post indicator shows location of fire hydrant if it is covered by snow.

Figure 6-32. Reflector set in the pavement provides quick indication of location of fire hydrant.

- In areas where significant snowfalls can be expected, verifying that some type of marker is in place to locate the hydrant if it is covered or hidden by snow from plows.
- Looking for any physical damage to the hydrant. If a hydrant has been struck by a vehicle or other object, pay special attention to your operational tests.
- Checking the overall condition of the hydrant. Does it need to be painted? Are the chains that secure the caps to the hydrant fastened securely and can the caps turn freely? Has the operating nut been damaged by the use of a pipe wrench or other unauthorized tool to open it?
- Making sure that the hydrant wrench you generally use fits all caps and the operating spindle.
- Removing all caps to verify that the metal has not oxidized and frozen the cap to the nipple.
- Checking all outlet threads for condition, lubrication, and tightness (It is common to see the nipples in the outlets loosen with abuse).

PRESSURE TEST

Before you open any hydrant, check whether the barrel is clear of any foreign material and make sure no water is present. Open the 4½″-outlet, and use a mirror and flashlight to physically examine the inside of the barrel. If there is water in the hydrant, either the hydrant was not closed completely after its last use, or its drain valve is not working properly.

With at least one of the caps removed, open the hydrant fully and allow water to flush any debris or foreign material out of the system to prevent damage to the pressure gauges. After you flush to clear up any discoloration, close the hydrant slowly to prevent water hammer in the system. Check to make sure that it drains after it has been closed.

Put a tapped hydrant cap, with a gauge attached, on one outlet of the hydrant to perform your pressure test. Tighten the caps on the other outlets and open the hydrant valve fully. After the air escapes, either through a bleeder valve on the tapped cap or through a loosened cap on one of the unused outlets, record the static reading on the gauge on your maintenance and inspection card for the hydrant you are testing. (A sample hydrant record form is shown in Figure 6-28.) While the hydrant is pressurized, inspect for leaks around the spindle, bonnet, nipples, caps, or ground flange. If water is leaking around one of the caps, replace the gasket. If you see water leaking around the operating nut, the packing may need to be adjusted or replaced.

FLOW TESTS

Immediately after you perform the pressure test, do a flow test for the hydrant's inspection record. Since you are checking the capacity of the individual hydrant rather than the system, install the cap gauge on one of the 2½″-outlets, and use a Pitot gauge on the other to compare the flow in GPM with the residual pressure at that rate of flow. By comparing your results from each hydrant inspection with the hydrant's original test, you can identify problems in the early stages, and correct them before major problems develop. Figure 6-14 shows a set of gauges designed specifically for hydrant testing. These gauges simplify the process as well as remove some of the variables introduced by the use of hand-held Pitot gauges.

Make sure that your flow test results are consistent with the color coding on the hydrant. If they are not, you will need to run additional tests. The tests should determine whether valves were left closed, whether modifications were made to the system that were not incorporated into the flow testing information, or whether system usage increased to the point that the expected fire flow is no longer available. This information should be available from the water department; if it is not there is poor communication between the water department and its inspectors.

After the flow test, close the hydrant completely and verify that all water drains from the barrel. Back off the operating nut far enough to remove the pressure from the thrust bearing or packing, one quarter turn or less, but not far enough to close the drain valve.

LUBRICATION AND REPAIR

After you have completed all inspections and tests, note any required repairs and refer them to the maintenance department. If any deficiencies are serious enough to take the hydrant out of service, mark it "Out of Service" and notify all stations and personnel that would need to use it. Revise any plans that include the defective hydrant and note alternate water supplies. Maintain a "hydrant-out-of-service" file, and devise a followup system to see that all repairs are made on a timely basis. If such a system is not in place, notify the responsible authorities.

If there is an obstruction in the drain holes and the hydrant is draining very slowly or not at all, flush them under pressure; this might eliminate the problem. Try opening the hydrant a few turns with all outlets capped, and allow the water that comes into the hydrant to escape through the drain for a few seconds. If you need more pressure, close the hydrant completely and attach a line from a pumper (using the proper fittings) to one of the 2½"-outlets. You can then use the water in the tank and the fire pump to apply pressure to the hydrant. However, limit the flow from the pumper to prevent washing out the drain field and damaging the hydrant. Try building the pressure on the pumper to 200 PSI, then slowly opening the gate valve on the outlet that is connected to the hydrant only far enough to allow the pressure on the individual line gauge to build up to 200 PSI. Apply this pressure only for a few seconds. If the obstruction is not removed upon initial pressure, consider this approach unsuccessful, and assume that the hydrant will have to be repaired.

MAKE SURE THAT THE MAIN HYDRANT VALVE IS COMPLETELY CLOSED BEFORE APPLYING PRESSURE TO THE BARREL FROM AN EXTERNAL SOURCE; THIS PREVENTS DAMAGE AND SYSTEM CONTAMINATION.

During all of these tests, the hydrant should be easy to open and close. If you apply lubricating oil to the valve stem and the packing each time the hydrant is inspected, it should make the hydrant easier to operate and maintain in good condition.

Before you put the caps back on the hydrant, check the threads; remove any rust or corrosion with a wire brush. Apply a light coating of a mixture of graphite and lubricating oil to the threads, and wipe off the excess. Check the retaining chains, where provided, to make sure they are free of excess paint, and that they hang straight, and will turn freely when you remove or install the caps.

FIRE DEPARTMENT RESPONSIBILITIES

The fire department should maintain complete records for each hydrant in a system. If the water department does the testing and maintenance, the fire department should receive a duplicate copy of all inspection reports and maintenance records. Report needed repairs to the water department in writing; make sure followup inspections are completed in a timely manner.

7. Suction Supply As A Water Source

OVERVIEW

FIRE HYDRANTS LOCATED ON approved water systems generally provide the best source of water for firefighting, but suction supplies offer a viable alternative. Many areas do not have a central water system. While some will have one eventually, others never will. Public water systems tend to lead to higher density development than is found in areas where water is supplied by individual wells. Planners frequently use the lack of a central water system as a tool to limit the spread of high density development. In these cases, other arrangements must be made for fire protection, and some form of suction supply is the best alternative. Suction sources also can be used to supplement inadequate public water systems, or to provide higher fire flows for special risks and target hazards.

FINDING A SOURCE

The primary requisite for a successful water supply operation that depends on drafting from a static (nonpressurized) source is a dependable, continuous water supply. The ISO's 1980 Fire Suppression Rating Schedule requires that available fire flow duration be 2 hours for a Needed Fire Flow (NFF) up to 2,500 GPM, and 3 hours for a NFF up to 3,500 GPM; higher flows must be sustained for longer periods of time. A source that has enough water in storage, a replenishment rate that exceeds the maximum flow requirements, or a combination of storage and flow meets the need. Based on your fire department's ability to transport water, a sufficient number of potential water sources and suction supply points should be identified to provide an adequate level of fire flow for the required period of time.

SOURCE CAPACITY

If you are taking the suction supply from a source of impounded water (e.g., pond, lake, cistern, or storage tank), estimate the average daily minimum storage by using the dimensions of the storage facility to determine the volume it contains when the water level is at its lowest point. Estimate the area covered by the body of water in square feet, then multiply this surface area by the average depth of the water, at its lowest point at the time of the day and season of the year, to determine the cubic feet of water in storage. If you multiply the cubic feet of water by 7.5 gallons, you can estimate the amount of water available for fire protection and use that total volume to find out how long the supply will last at the needed rate of flow. Use the formula:

$$L \times W \times D \times 7.5 = C$$

where

L = average length of the body of water
W = average width of the body of water
D = average overall depth of the body of water
C = capacity of the body of water in gallons

For example, a farm pond averages 100 feet in length, 70 feet in width, and 6 feet deep when it is at its lowest normal level.

$$200 \times 70 \times 6 \times 7.5 = 630,000 \text{ gallons}$$

This pond's minimum daily storage would be 630,000 gallons. It is used to provide fire protection for a storage building which has been estimated to require a NFF of 2,500 GPM based on the fire flow calculations illustrated in Chapter 3: Risk Analysis in Fire Protection Planning. Use the following formula to find out if it meets the minimum storage:

$$\frac{C}{NFF \times 60} = D$$

$$\frac{630{,}000}{2{,}500 \times 60} = D$$

$$\frac{630{,}000}{150{,}000} = 4.2$$

You can see from the calculations that this pond could provide 2,500 GPM for approximately 4.2 hours before it empties, thereby meeting the ISO's NFF requirements. If you are trying to determine whether a particular body of water can supply the NFF, you must consider the drafting capability of the pumpers you will use. Since the maximum GPM a pumper can supply is limited by its pump capacity, the length of suction hose needed to reach the water, and the lift height, in this case, at least two pumpers, with a total capacity of 2,500 GPM, would have to respond and be able to get close enough to the water to deliver their rated capacity in order to provide the NFF to this structure.

SOURCE RECHARGE OR FLOW RATE

You can draft from a rapidly flowing stream, river, or other source instead of a static body of water. Without water held in storage to provide a reserve, the maximum flow rate is limited by the rate of flow in the stream. The most accurate way to estimate the maximum flow rate is to use a weir, inserted into a dam, as a measuring gauge (see Figure 7-1).

Many weirs are calibrated with a GPM figure marked on them, while others are marked with a gallons per day (GPD) figure. You can convert this number to gallons per minute by dividing the GPD by 1,440. Since all of the water is forced over the weir by the obstruction in the stream, the flow reading should be accurate, and provide a dependable measure of the flow available for fire protection.

If a stream has an irregular waterway, it is difficult to construct a dam that blocks the movement of water. Some of the same materials designed to contain hazardous materials spills can be used to form a reasonably good seal, and direct most of the flow through the weir in the dam. Alternatively, you can estimate the flow in a free-flowing portion of the stream.

First, select a section of stream bed which is relatively unobstructed, and has an even flow throughout.

Next, drop a float in the stream and mark a reference point along the bank. Observe the float for a period of 10 seconds.

Mark the distance traveled by the float. Measure the distance in feet.

Then, estimate the cross-sectional area of the stream in square feet. (See Figure 7-2, for an example) If the stream's average depth is 2 feet, and the width 6 feet, the cross-sectional area is 12 feet. Multiply the area of the moving water by the distance in feet the float traveled in ten seconds to estimate the amount of water that is passing the reference point. Divide the distance the float traveled in 10 seconds by 10 to determine water velocity in cubic feet per second. Multiply the cubic feet per second by 60 to determine the cubic feet per minute flowing at that point.

Last, multiply the volume of water in cubic feet by 7.5 to determine the stream's gallons per minute flow.

To make these calculations, use:

$$D \times W \times V \times 60 \times 7.5 = GPM$$

where
- D = average depth of the stream
- W = average width of the stream in the area where the flow is being measured
- V = velocity of the water in feet per second. This measurement can be calculated by dividing the distance the float traveled by the number of seconds from the start of the test to the end
- 60 converts the cubic feet per second to cubic feet per minute
- 7.5 converts the measurement in cubic feet per minute to gallons per minute flow

For example, assume that a float in a stream travels 20 feet in 10 seconds. If the stream's average depth is 2 feet, and its average width is 10 feet

$$2 \times 10 \times (20/10) \times 60 \times 7.5 = 18,000 \text{ GPM}$$

If you are performing a flow test in a stream of moving water, the rate of flow varies from the edges of the stream into the center of the chan-

Suction Supply as a Water Source 169

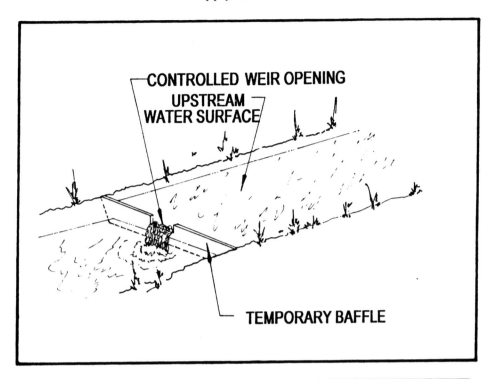

Figure 7-1. Calibrated weir inserted in temporary dam to measure flow in a stream.

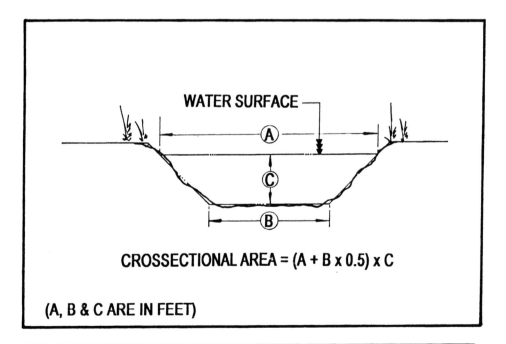

Figure 7-2. A method of estimating the area of a stream for calculating flow.

nel. The more irregular the banks of the stream, the greater the variation. A float generally moves to the area of most rapid flow, so the average flow is somewhat less than the GPM figure shown in the calculations above. You can derive a more conservative estimate by taking 75% of the measured flow as the safe yield from the stream. In the example shown, 75% of 18,000 GPM yields an available fire flow of 13,500 GPM, calculated

$$18{,}000 \times .75 = 13{,}500 \text{ GPM}$$

Any estimate of flow is only a rough approximation of how much water the stream could produce at the time of the test. Seasonal variations, average rainfall, and the water other users take all can affect the amount of flow at any given time. You must consider all of these variables when you evaluate the water source's potential for fire protection purposes.

CERTIFICATION OF DEPENDABILITY

If you are preplanning for specific risks, and preparing a comprehensive fire protection plan, all you need is an estimate of a water source's capacity. If you are using the water supply point to qualify for an ISO rating, a registered professional engineer needs to certify the minimum daily storage, or the maximum rate of flow that would be available in a 50-year drought. If the United States Department of Agriculture's Soil Conservation Service has information about the area, it also may be able to provide the required certification.

Privately owned or controlled water sources must have both an engineer's certification of proposed reliability, and a written agreement from the property owner that allows the fire department to use the water supply for training and emergencies. Any improvements necessary to make the water source usable should be detailed in the agreement, specifically who will see that improvements are made and how they will be paid for. Arrangements for notifying the fire department of any change that will affect the availability of the source for water supply must be included in the agreement. The fire department must have guaranteed access to the property for any required maintenance and testing. One mechanism for providing access is to add a standard fire department lock to the property owner's lock on a chain or a gate. By keying all fire department water supply points to the same lock, a pumper with the key can access any drafting location, even on a mutual aid basis.

Suction Supply as a Water Source 171

Figure 7-3. Normal level in lake used for water supply.

Figure 7-4. Lake is unusable when level drops.

Figure 7-5. Addition of a "fire department lock" provides access to water supply point.

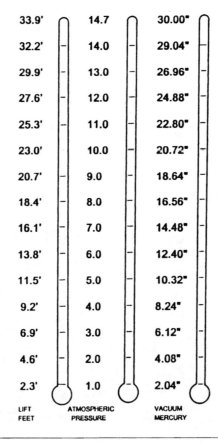

Figure 7-6. Chart showing the relationship between atmospheric pressure, feet of lift, and inches of mercury (vacuum) when drafting.

EXCESSIVE LIFT

One of the strictest limitations to the amount of water that a pumper can supply from draft is the lift from the surface of the water to the eye of the impeller inside the pump. A fire department pumper drafts by removing air from the pump and suction hose, thereby creating a pressure differential. The atmospheric pressure acts on the surface of the water to force it through the suction hose and into the pump. This pressure differential must be great enough to overcome the head pressure of the water as it fills the suction hose during priming. When the level of the water in the suction hose reaches the eye of the fire pump impeller, it is considered primed and ready to draft water from the source. When the pump is primed and discharging water, the flow through the pump creates enough vacuum on the intake to keep it filled. Figure 7-6 shows the relationship between atmospheric pressure, vacuum on the intake of the pump, and the equivalent feet in lift.

The water's weight creates a head pressure of .434 PSI for every foot of lift. When the atmospheric pressure is 14.7 PSI, the normal pressure at sea level, the maximum theoretical lift would be 33.9 feet. To reach this lift requires a perfect vacuum, i.e., 30 inches of mercury on a vacuum gauge; this is impossible for fire department pumpers. Among the factors that limit maximum lift for a fire pump are

1. The average atmospheric pressure decreases with elevation. Approximately .5 PSI of atmospheric pressure is lost for each 1,000 feet of elevation above sea level.
2. No fire pump can create a perfect vacuum. According to NFPA 1901, *Standard for Pumper Fire Apparatus*, the primer must be able to develop a vacuum of 22 inches within 30 seconds.

Given these limitations, most pumpers cannot be primed when the lift from the surface of the water to the eye of the impeller is more than 25 feet. Once the pump is primed, and the water begins to move, you face other losses.

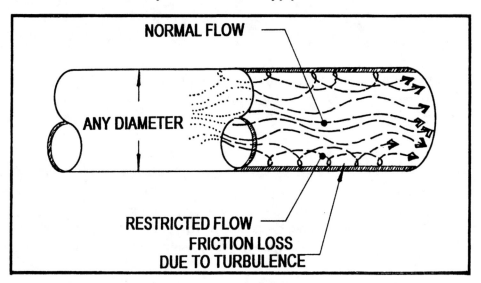

Figure 7-7. Friction loss occurs as water moves through a fire hose.

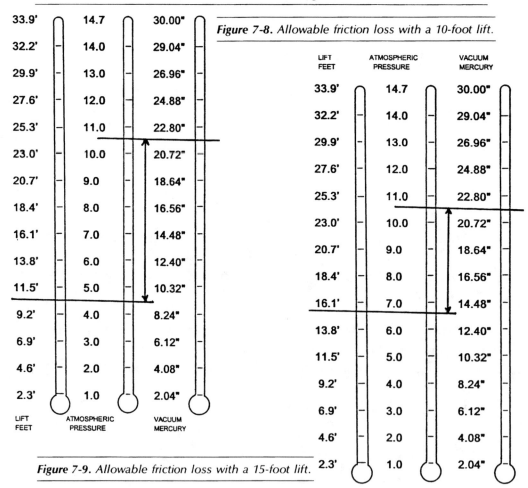

Figure 7-8. Allowable friction loss with a 10-foot lift.

Figure 7-9. Allowable friction loss with a 15-foot lift.

FRICTION LOSS

As water flows through a pump, pressure loss occurs in the suction hose because there is friction against the hose's interior surface. In Figure 7-7, you can see friction's effect on water movement through a fire hose, and the resulting opposition to flow. Figure 7-6 shows that atmospheric pressure limits the maximum pressure available to force water into a fire pump. Some of this pressure is used to overcome head pressure; only the remainder is available for friction loss. The higher the lift, the more head pressure there is to overcome, and the lower the amount of tolerable friction loss. For example, Figure 7-8 shows that with 22 inches of vacuum, and a 10-foot lift, approximately 6 PSI of atmospheric pressure is available for friction loss in the suction hose. If you increase the lift by 5 feet to 15 feet, as is shown in Figure 7-9, only about 4 PSI is left for friction loss. To minimize the friction loss increase the size of the suction hose to supply higher capacity pumps. Table 7.1 gives estimated maximum flows from a given pumper with changes in lift.

Table 7.1. Estimated Maximum Flow With Changes in Lift and Pump Capacity

Feet of Lift	750 GPM		1,000 GPM		1,250 GPM		1,500 GPM	
	GPM	%CHG	GPM	%CHG	GPM	%CHG	GPM	%CHG
4	870	+16%	1,160	+15%	1,435	+15%	1,735	+15%
6	830	+10%	1,110	+10%	1,375	+10%	1,660	+10%
8	790	+5%	1,055	+5%	1,310	+5%	1,575	+5%
10	750		1,000		1,250		1,500	
12	700	-6%	935	-6%	1,175	-6%	1,410	-6%
14	650	-13%	870	-13%	1,100	-12%	1,325	-12%
16	585	-22%	790	-21%	1,020	-18%	1,225	-18%
18	495	-34%	670	-33%	900	-28%	1,085	-28%
20	425	-43%	590	-41%	790	-37%	955	-36%

NOTE: 1. Percentage changes and pump capacity calculations are only approximations intended for fireground use.
2. Table 7.1 is based on: 4½"-suction hose on a 750-GPM pump; 5"-suction hose on a 1,000-GPM pump; and 6"-hose for the 1,250- and 1,500-GPM pumps.
3. These estimates assume 20 feet of suction hose up to a 16-foot lift with a third section added for 18 and 20 feet.
4. A pump in good condition will supply water at a lift higher than 20 feet, but the maximum flow would not be much more than 50% of the pump's rated capacity.

Figure 7-10. Low-level bridge allows pumper to supply maximum flow from draft.

CAVITATION

When a pump is primed initially, the vacuum reading on the compound gauge at its intake is proportional to the lift. Each foot of lift amounts to .88 inches of vacuum (as illustrated by Figure 7-6), or for fireground estimations, one inch per foot. As water begins to flow, losses in the suction hose and strainer add to the vacuum or negative pressure, reading at the intake. As flow increases, losses in the system increase, and the vacuum reading goes up. If the flow continues to increase until the negative pressure at the inlet to the pump is equal to the atmospheric pressure, water will continue to enter the pump, but the flow rate will not increase. If additional lines are put into service, the pump attempts to discharge water faster than it is coming in, and the pump "runs away" from the water. At this point, cavities of water vapor form in the intake, or low pressure side of the pump. As these cavities pass through the pump and go through the high pressure areas inside the housing, they will collapse with some violence, causing fluctuations in the flow, variations on the discharge pressure gauge, and an uneven fire stream as it leaves the nozzle. More importantly, the violence of the implosion as these cavities collapse causes damage to the fire pump. Indications that a pump is operating beyond the point of cavitation include

- a high vacuum or negative pressure reading on the compound gauge at the intake of the pump;

- a "rattling" noise as water passes through the pump;
- pressure gauge fluctuations;
- fluctuations in the fire stream as it leaves the nozzle; and
- an increase in the throttle setting and engine RPM does not produce a corresponding increase in the discharge pressure.

Cavitation may result in a fire pump that is operating from draft when additional hoselines are put into service. Because many of these indications are not obvious, a danger is that the pump operator might not realize it is happening. Pump operators must pay close attention to the gauges and instruments so that they can make the adjustments necessary to keep the pump from cavitating and maintain a dependable supply to the attack lines.

Water supply officers must be aware of how many lines can be supplied from a drafting operation so that they can limit demands on the pumper. At the same time, it is the pump operator's responsibility to tell the water supply officer how much additional flow can be supplied. You can estimate a pump's remaining capacity by observing the vacuum reading on the intake gauge. Most pumpers experience cavitation when their compound gauges read between 20 and 25 inches. Friction loss (and the vacuum reading) increases exponentially. If the negative pressure created by a flow of 500 GPM registers 3 inches of vacuum, increasing the flow to 1,000 GPM would change the vacuum reading to 12 inches. If a vacuum reading of 10 inches is required to overcome the head pressure created by the lift, the reading on the compound gauge would be 22 inches at this point. This would be a typical vacuum reading for a 1,000-GPM pump operating at its rated capacity from a 10-foot lift, with 5″-suction hose; therefore, any time the vacuum reading on the intake exceeds 15 inches, the pump operator should tell the water supply officer that the pump is approaching its maximum flow capability. Find another source (or add a second supply pumper on draft) if a significant increase in the needed fire flow is anticipated.

Trying to read a compound gauge accurately is difficult and problematic in drafting operations. Most pumpers do not have separate vacuum gauges, and the vacuum portion of a compound gauge covers less than five % of the total scale. Given the typical calibration variations of pump panel gauges, it is difficult to determine an exact vacuum reading. In Chapter 13: Specifications for Mobile Water Supply Apparatus, you'll read a suggestion that pumpers should have separate vacuum gauges on their intakes to make it possible to obtain accurate readings. A separate vacuum gauge both makes it easier to recognize the point of cavitation, and indicates when a pump is approaching it, allowing the water supply officer time to make provisions for additional flow before it is needed.

The most accurate, positive indication of fire pump cavitation is when an increase in engine RPMs does not yield a corresponding increase in discharge pressure. When the maximum amount of water supplied to a pump equals water discharging, increasing the impeller's speed of rotation supplies no additional water to the discharge, and pressure does not increase. If an operator suspects a pump is cavitating, he or she should decrease the throttle until discharge pressure drops, then increase it until the pressure gauge shows the maximum pressure reading. There should be no increases above that point.

WARNING: IF YOU ARE OPERATING AT MAXIMUM FLOW, ANY CHANGES IN NOZZLE SETTINGS OR ADDITIONAL LINES MAY CAUSE PUMP CAVITATION.

MINIMUM DEPTH

A good suction source has enough depth to maintain a minimum of 18 inches of water over the strainer. A pumper operating at or near its rated capacity draws water into the strainer at a very rapid rate, forming a whirlpool. If the whirlpool effect is strong enough, air is drawn into the pump, causing it to lose its prime.

If your supply source is a pond or stream with a sandy bottom or one covered with loose dirt and debris, you need to position the strainer so that foreign objects and material, which could damage the pump, will not be drawn into it. Some fire departments put a bucket over the end of the strainer to minimize pump damage. Others use a ladder or ropes tied from the strainer to a high point on the pumper; this supports the end of the suction hose to keep it at least 18 inches higher than the bottom of the source. The strainer also can be supported by floating with an empty foam can, etc.

When a swiftly flowing stream is not deep enough to draft effectively you can construct a dam using available materials. Since many pumpers carry salvage covers or other materials, use these to block the flow of water and increase the depth at the proposed drafting location. Ladders, pike poles, or charged hoselines are options for building a framework to keep dam obstructions in place. A line of large diameter hose can be used this way. In Figure 7-15, you can see rocks, pieces of wood, or other materials found on site used for this purpose in an emergency. Another option is to dig out the bottom to increase depth.

Figure 7-11. A bucket will make a usable float for a barrel strainer and keep it off the bottom.

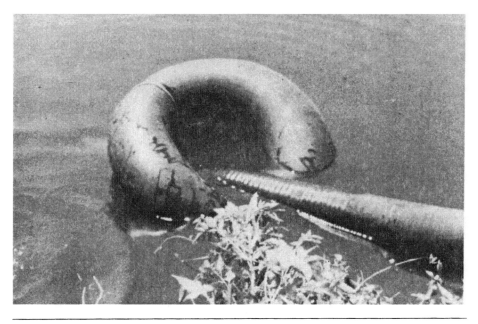

Figure 7-12. To save space, a truck innertube can be carried deflated, and inflated from the air brake system to serve as a float for a strainer.

Figure 7-13. *A roll of plastic can be used to construct a dam to create enough depth to draft from a shallow stream.*

Figure 7-14. *A salvage cover makes a good dam for drafting.*

A much better solution to the problem of inadequate depth is to use a floating strainer, which floats on the surface and takes in water from the strainer's underside. Because water is drawn into the strainer from all directions, it has to pass under the edges of the float, thus minimizing the whirlpool effect. Since the strainer floats on the surface, it is less likely that dirt and debris will be drawn into the pump from the bottom of the water source. Potential draft pumpers should carry floating strainers (this eliminates the need to carry a barrel strainer). The specifications for all new fire apparatus should require floating strainers instead of conventional barrel-type strainers.

WATER QUALITY

While you can use any type of water in an emergency, pumping contaminated water can damage a fire pump over time. One pump mechanic said, "I wouldn't pump any water I wouldn't drink." This might be a good rule to follow in the best of worlds but in the real world, where most rural departments operate, this luxury is not permitted. What is important is to take precautions when you draft water that you would not drink.

DIRTY WATER

If you pump water that contains sediment and debris, large particles can enter the intake and lodge in narrow passages in the impeller or other areas of the pump. Since obstructions can prevent a pump from delivering its rated capacity, it may have to be dismantled to be cleared. Such repairs can be expensive. Both the strainer on the end of the suction hose and the strainer at the inlet to the pump must be maintained. Openings in these strainers should be small enough to keep out dangerous objects.

If there is a significant amount of vegetation or floating debris in the water, it can accumulate at the strainer, blocking some of the holes and obstructing the water's movement. This effect is most pronounced when the flow through the pump nears its maximum, and the velocity of the water being drawn into the strainer increases. With the strainer obstructed, friction loss increases; the pump might cavitate, even if the lift is not excessive and the pump has not reached its maximum capacity. When the vacuum reading increases with no increase in flow through the pump, you need to check the strainer for blockage. It might

Figure 7-15. There are generally enough materials on site to construct a dam for drafting.

Figure 7-16. A floating strainer will enable a pumper to draft from as little as twelve inches of water.

be necessary for someone to wade out into the water and clean the strainer periodically. If you are drafting from a stream, construct a deflector from stones, wood, or other available materials, to force floating debris to bypass the strainer. If nothing else works, try limiting the flow through the pump to reduce the water's velocity and minimize the accumulation of vegetation and debris at the strainer.

Sand or sediment suspended in water causes long-term pump damage. Close tolerances inside the pump between the wear rings and the impeller, between the packing and the impeller shaft, and inside the various accessories used with fire pumps, cause dirt in the water to act as an abrasive, resulting in excessive wear. There is little you can do to stop this other than operating at the minimum possible pressure, and keeping the pump's RPMs as low as possible.

Contaminated Water

In many areas water used for fire protection contains contaminants, one of which is the salt water available in areas near oceans. Sometimes surface water has chemicals from mining or manufacturing activities.

The only way to minimize the effects of contaminated or dirty water is to back-flush your pump as soon as possible after you use it. When a source of clean, pure water is available, connect a line under pressure to the highest discharge outlet on the pump. Open all suction openings and let water flow freely through the pump in the direction opposite the normal movement of water. Back-flushing should dislodge any debris caught inside the pump and flush it out of the intake. After you back-flush, connect a clean water supply to one of the intakes, and use the pump to build up normal operating pressure. Operate all accessories, including the relief valve and primer, to allow the clean water to flush out any sediment or contaminated water that may have been trapped in the lines and to flush the packing rings. If you flush the pump thoroughly each time it has pumped contaminated or dirty water, you can minimize damage and extend the life of the pump.

ACCESSIBILITY

A good suction supply point should be accessible to pumpers at all times of the year and in all kinds of weather conditions. To perform at maximum capacity, a pumper must be close enough to the water source to draft through 20 feet of suction hose, with a lift of no more than 10 feet. There also must be enough room between the pumper and the edge of the water to provide a safe working position for the pump operator. In general, all-weather roads should be constructed to within 6 feet of the water to provide pumpers greater access to water sources.

Boat ramps provide pumpers good safe access for drafting, but generally are not very wide. It sometimes is difficult, if not impossible, to position a pumper to connect the hard suction hose to the intake at the side of a traditional midship-mounted fire pump. Front-mount pumps are much better suited for drafting when your maneuvering room is limited. Some midship-mounted pumps also have a front suction intake, but its primary purpose is for use on pressurized fire hydrants. Since it is difficult to get the large pipe required for drafting from the pump to the front of the vehicle, many of these lines have sharp bends and other fittings that increase friction loss. Many 1,000- and 1,250-GPM pumps have front suction lines with 5″-, or smaller, pipe from the pump intake manifold to the front of the vehicle. So, you seldom see a midship pumper that can supply more than 1,000 GPM drafting through its front suction (many are limited to much less than 1,000 GPM). Drafting with a front suction is problematic because there generally are high points in the line where air pockets are likely to develop as water enters the pump, making it difficult to prime. If a pumper is configured to draft through a front suction, consider a requirement to install additional lines to the primer at all high points in the suction line and specify a minimum flow rate when drafting from a 10-foot lift.

Figure 7-17. Three or more sections of hard suction hose can be used to reach a water source that would be inaccessible, but with a reduced maximum flow.

If you use a third 10-foot section of suction hose your pumper can draft from a greater distance and give more flexibility; this will, however, reduce the maximum flow available.

DRAFTING HYDRANTS

When you cannot position a pumper close enough to a water source to draft directly, you can use drafting hydrants to get a reliable water supply. Drafting hydrants would be beneficial where fences, drainage ditches, or other topographical features bar your access (see Figure 7-18). Ponds frequently are located in pastures where confined animals make access difficult. In some cases, sloping pond banks are too steep for access. Ground too soft or too wet to support a pumper's weight between the road and the water source, weather conditions that block access, and locations where property owners are not willing to permit a driveway or access road for aesthetic reasons all are situations for which installing a drafting hydrant would be beneficial. Regardless, drafting hydrants should supply enough water to operate a pumper at its maximum rated capacity.

A drafting hydrant has three parts:

- A pipe from the drafting location to the water source;
- An adapter to connect the standard hard suction hose from a pumper to the pipe extending into the water source. (A strainer should be a part of this adapter); and
- A strainer on the source end of the pipe to prevent debris from being drawn into the drafting hydrant system.

There are different types of drafting hydrants. "Wet hydrants" are located below the level of the surface of the water. Wet hydrant barrels are filled by gravity's force; they require a valve to shut off the flow while a suction hose is being connected. When the valve is opened, the pump fills with water without the primer. What differentiates a "wet hydrant" from a "pressurized hydrant" is that a pumper has to draft from a wet hydrant because a wet hydrant has less than 20 PSI static pressure, not enough to provide an adequate supply to the pumper. The vacuum the fire pump generates is used to overcome friction loss in the pipe rather than lift. Figure 7-20 shows one type of wet hydrant installation that uses a pond's natural elevation. In Figure 7-21, you can see a fire-department-installed 14,000-gallon storage tank, with a 6"-line to a standard fire hydrant. The static pressure at this hydrant is 8 PSI, and the fire department can get a flow of approximately 450 GPM

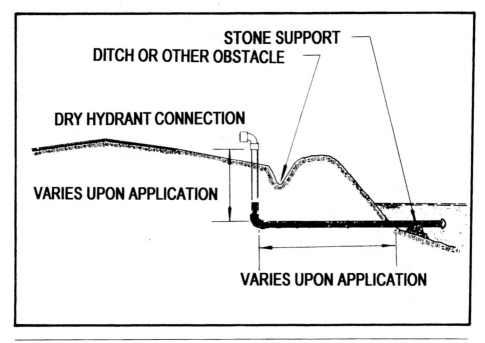

Figure 7-18. A drafting hydrant can be used to gain access to an inaccessible source.

Figure 7-19. Dry hydrant set in concrete to gain access to a pond.

Suction Supply as a Water Source 187

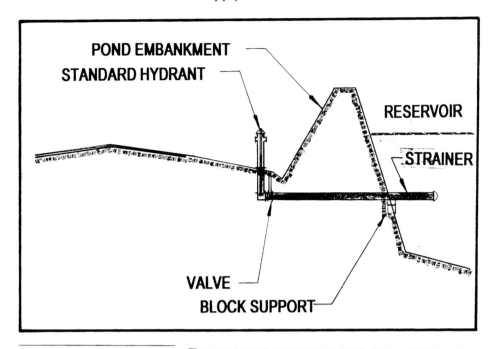

Figure 7-20. When the water level in the source is higher than the drafting hydrant, the line will be full of water at all times.

Figure 7-21. Overhead storage tank maintains a pressure of 8 PSI at the drafting hydrant.

by head pressure alone. When this department uses a pumper to draft from the hydrant, flows in excess of 1,100 GPM are possible.

A "dry hydrant" is more typical of drafting hydrants because the pumper connection is located above the surface of the water. Since the hydrant is normally dry, the primer on the fire pump is used to bleed air from the pipe, create a vacuum, and then fill the hydrant with water. In Figure 7-18 the diagram shows the installation of a typical dry hydrant.

SYSTEM DESIGN

What limits the maximum flow rate of a drafting hydrant are static losses, such as the lift from the water to the fire pump, atmospheric pressure, and water temperature, all results of natural causes. Static pressure loss remains the same regardless of the rate of flow from the hydrant. A dynamic pressure loss also is incurred as water moves through the drafting hydrant and associated piping. The amount of dynamic pressure loss is proportional to the velocity of the water as it travels through the waterway and is dependent on the flow rate and pipe size.

Most drafting hydrants use 6"-pipe to get water from the static source to the pumper. Since atmospheric pressure is used to force the water into the pump, the flow can increase only until the losses in the system equal the atmospheric pressure on the water's surface. Six-inch pipe is generally adequate for wet hydrants. Since the water source is above the drafting hydrant, pressure gain comes from the difference in elevation. With no lift, the only losses to be overcome by atmospheric pressure result from water movement through the system. With a dry hydrant, the atmospheric pressure has to overcome both the head pressure from the lift and the friction loss in the system. As a result, a typical dry hydrant, with 6"-pipe, generally can supply between 600 and 1,000 GPM, not always enough to sustain needed fire flow or to fill tankers fast enough to operate a high-capacity water shuttle. If you decide to go to the trouble and expense of installing a dry hydrant, make sure it can supply the full capacity of the pumpers it will serve.

Some of the agencies responsible for designing and installing dry hydrants use the ISO's Fire Department Water Supply criteria as a standard for the required flow. The ISO requires at least a 250-GPM flow to qualify, but awards credit for additional flow that can be developed within 15 minutes of arrival on an incident scene. A common misunderstanding about setting standards for dry hydrant installation is that

Figure 7-22. A wet drafting hydrant still requires a pumper to draft to attain a flow of 1,000 GPM or more.

Figure 7-23. Draft site with a dry hydrant for one pumper to draft and a trough to make it easier for a second pumper to put hard suction hose in the water.

a dry hydrant that can supply 250 GPM adequately meets the ISO standard, when, in reality, you have to fill tankers at a rate approaching 1,000 GPM to maintain 250 GPM by water shuttle. Fire departments need to verify that designers and installers of dry hydrants use a minimum flow of 1,000-GPM as the minimum standard.

You need pipe larger than 6" to install a dry hydrant that will reliably supply more than 1,000 GPM when the lift is more than 10 feet. Too much of the available atmospheric pressure is used to overcome the head pressure and the maximum flow through the 6"-pipe is limited by the remaining pressure available. When the distance to the water source from the proposed drafting location is greater than 50 feet, the friction loss in the 6"-pipe generally limits the maximum flow to something under 1,000 GPM.

In general, use the following criteria to determine required pipe size:

Six-inch pipe: No more than a 10-foot lift and 50 feet of lateral pipe.
Eight-inch pipe: No more than a 15-foot lift and up to 50 feet of lateral pipe.
Ten-inch pipe: Up to a 15-foot lift with as much as 400 feet of lateral pipe.

You can use any type of pipe for a drafting hydrant. Early drafting hydrants used standard cast iron water pipe, not the ideal choice because it is difficult to make airtight connections with it and it is susceptible to rust and corrosion, significantly increasing friction loss and reducing the maximum flow.

Most modern dry hydrants use Schedule 40 PVC pipe. PVC pipe has the advantages of a smooth interior to minimize friction loss, is easily assembled with pipe cement to prevent air leaks, and doesn't deteriorate with age. One drawback is that Schedule 40 PVC pipe is adversely affected by sunlight. The sun's ultraviolet rays eventually will cause the pipe to become brittle and lose its resiliency. The portion of the hydrant that extends above the ground should be painted to help protect it from the sun.

EXCESSIVE LIFT

One of the recommendations in Calvert County, Maryland's Water Supply Master Plan, adopted in 1989, was to install dry hydrants in a number of locations. Ultimately dry hydrants were installed in three locations. Since the lift on two of the hydrants was greater than 10 feet, 8″-PVC pipe, reduced to a standard dry hydrant with 6″-NST threads on the pumper connection, was used. The results of the initial tests of these hydrants are listed in Table 7.2.

Table 7.2. Dry Hydrant Test Results (Calvert County, Md.)

Location	Length Pipe	Size Pipe	Lift	Priming Time	GPM flow
Victoria Station	71′	8″	13′	48 sec.	1,043
Lake Karylbrook	73′	8″	16′	45 sec.	1,087
Wohlgemuth Rd.	40′	8″	8′	35 sec.	1,122

INSTALLATION

In Calvert County's case, the hydrants were needed to provide a flow of at least 1,000 GPM for tanker loading. To obtain the maximum flow, they were installed initially with 6″-pumper connections with NST thread. Forty-five-degree bends instead of 90-degree elbows (similar to the method of installation shown in Figure 7-24) were used to minimize friction loss.

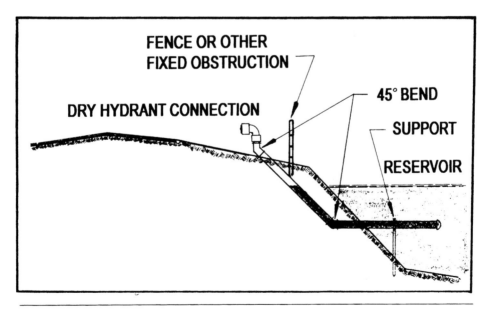

Figure 7-24. Friction loss in a dry hydrant system can be reduced by using 45-degree bends in the line.

Figure 7-25. Left, a bed of gravel in the trench before a dry hydrant is installed will prevent damage to the line and potential air leaks.

Figure 7-26. Below, the pipe should be assembled before the dry hydrant is lowered into the trench.

Suction Supply as a Water Source 193

To provide a margin of safety, the hydrants were installed with the strainer on the end of the pipe four feet below the surface of the water versus the ISO's standard three-foot requirement. Then, if the water level in the pond were to drop during a dry season, the dry hydrant still could be used to provide its expected flow rate. You could install a dry hydrant less than three feet below the surface in a shallow pond, but it would have to be deep enough to keep the lateral section of pipe below the frost line. The lateral pipe's entire length should be lower than the water's surface; it will be full of water when the hydrant is not in use.

One option is to install a hydrant with 45-degree bends instead of 90-degree bends to reduce friction loss. If this option is used, a portion of the lateral/riser is not filled with water when the pumper connects to it as shown in Figure 7-24. The fire pump's priming device will have to evacuate all of the air from the dry section before it fills with water and the pump becomes primed. You need to extend priming time to evacuate additional air from the line. Table 7.2's test results show that all three hydrants tested required more than the NFPA-specified 30-second maximum time to prime a pump through 20 feet of suction hose. Pump manufacturers recommend against operating a primer dry for more than 60 seconds. If sections of dry pipe in a dry hydrant installation are too long, it could take longer than 60 seconds to prime the pump.

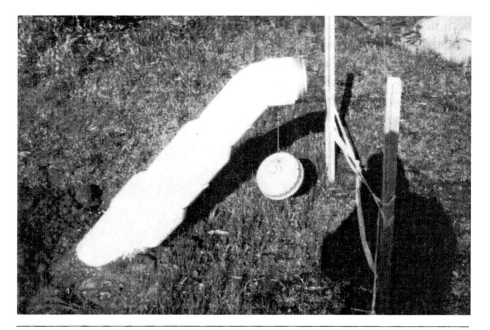

Figure 7-27. For maximum flow, 10"-pipe, reduced to 8", then to 6" at the pumper outlet should be used.

A typical fire pump's priming device is designed to remove the air from 20 feet of 6''-hard suction hose within 30 seconds. When dry, this length of hose contains approximately four cubic feet of air. To use a dry hydrant, you need at least one 10-foot section of hard suction hose from the fire pump to the dry hydrant. If the riser and the dry hydrant total less than 10 feet of 6''-pipe, it should be no more difficult to prime than to draft through 20 feet of suction hose. Conversely, if the installation includes more than ten feet of dry pipe, your priming time will lengthen proportionately.

The priming problem is aggravated when 8''- or 10''-pipe is used to install the dry hydrant. Instead of 2 cubic feet of air in each 10-foot section, 8''-pipe has approximately 3.5 cubic feet per 10-foot section, while a corresponding section of 10''-pipe would contain almost 5.5 cubic feet of air. Given the results in Table 7.2, and other tests, you should have no more than 20 feet of dry pipe above the water level with 6''-pipe, 15 feet with 8''-pipe, or 10-feet when 10''-pipe is required.

Dry hydrants are best installed in ponds that have been drained temporarily, and caps should be installed on the drafting connection to protect the threads. With a cap on the outlet, the dry hydrant is airtight and, as the pond fills with water, air cannot escape from the pipe. There often is a tendency for the pipe to float, so the strainer must be fastened securely or weighted down. You can keep the lateral pipe from floating by loosening the hydrant cap enough to allow air to enter or escape from the pipe as the pond level changes. Keep in mind, however, that vandals could remove the cap and put debris into the hydrant. Consider, instead, drilling a small vent hole in the cap to prevent vacuum or pressure buildup when water enters or flows out of the pipe when the pond level changes.

Figure 7-28. *A small hole in the cap will allow the air to escape and the level in the dry hydrant pipe to change with the level in the source.*

CONNECTING SUCTION HOSE

Your hydrant's drafting outlet should be approximately two feet above the ground. When the hydrant end of the suction hose is slightly lower than the pump's intake, the hose is more easily connected and there is less stress on the hydrant. This setup also tends to eliminate the air pocket at the high end of the dry hydrant because the hydrant end of the hose is somewhat lower than the pumper's steamer connection. Some type of support in the middle of the suction hose helps to minimize the stress on the PVC pipe and protect the dry hydrant from damage. Departments that use dry hydrants on a regular basis have devised various types of braces or stands to support the suction base and to ease the strain on the hydrant when the suction hose fills with water.

It is very difficult to connect conventional hard suction hose from the suction inlet on the pumper to the dry hydrant. You must maneuver the pumper precisely, which requires at least two people in addition to the driver, as well as a significant amount of time. Lightweight, flexible PVC suction hose is much easier to use because you simply position the pumper close to the hydrant. With the hose's flexibility and lightweight construction, one person can make the connection, important when a fill site is needed to load tankers in the early stages of a fire and staffing is at a premium.

Calvert County's dry hydrants initially were installed with 6''-outlets to ensure maximum flow. To use these outlets, their pumpers had to be equipped with double female adapters, with 6''-couplings on one end; the other ends were sized to fit the suction hose on the pumper. It was suggested that the dry hydrants be changed to 4½''-fittings with NST threads, identical to a standard pressurized fire hydrant. Since most pumpers are equipped with adapters to connect to standard fire hydrants, these same fittings can be used to connect to dry hydrants, assuming the couplings are compatible. Calvert County's original intent was to purchase dual 6''-double female adapters for each of its fire departments so that when mutual aid pumpers were assigned to set up a draft and fill tankers, they would have the fittings to make the connection. Instead of purchasing adapters, they decided to change the 6''-male connections at the hydrant to standard 4½''-threads. Using a standard 4½''-fitting for the dry hydrant would enable any pumper to connect to it and fill tankers with the equipment it normally carries.

Before Calvert County's 4½''-dry hydrants were installed, tests were conducted at both locations to determine maximum flow with the 6″-hydrants in place. The tests were repeated after the dry hydrants were changed to 4½″-fittings on the end of the 6″-pipe. The results of these tests are shown in Table 7.3.

Table 7.3.

Dry Hydrant	6″-Hydrant			4½″-Hydrant		
	Pitot	Tip	Flow	Pitot	Tip	Flow
Lake Karylbrook	48	2¼	1,043	52	2¼	1,087
Victoria Station	46	2¼	1,021	48	2¼	1,043

Both sets of tests were made at the same location using the same pumper, suction hose, and test equipment. Even allowing for a difference in atmospheric pressure between the two tests, it is difficult to account for the higher flow with the smaller threads and the adapter versus the initial test with 6″-suction all the way. The results indicate little, if any loss in maximum flow with a 4½″- versus 6″-dry hydrant.

EXCESSIVE DISTANCE TO THE SOURCE

The Bostian Heights Fire Department, located near Salisbury, North Carolina, was having difficulty supplying the needed fire flow for two large risks in its response area. While both risks had large ponds available, their distance from the road made them inaccessible for drafting.

Two large churches located in a rural area, approximately two miles apart, constituted the greatest threat. There was one large pond that could provide a dependable water supply, but it was nearly 400 feet off the road. Building an access road on the marshy, unstable ground was ruled out. The other pond, located on a farm with a number of agricultural type buildings, was more than 100 feet from the road, and had a lift of twelve feet. Fire department personnel decided to install dry hydrants, with 10″-Schedule 40 PVC pipe to minimize friction loss, in both locations. The 10″-pipe was used in an effort to provide a flow greater than 1,000 GPM through the extremely long suction line near the churches, and for the combination of a high lift and a long distance to the pond on the other.

A 1,500-GPM pumper was used to make maximum flow tests at both of these hydrants, and a master stream device was used with a Pitot gauge to determine flow. See Table 7.4 for the results.

Table 7.4. Dry Hydrant Test Results (Salisbury, N.C.)

Location	Flow Meter	Length Pipe	Size Pipe	Lift	Priming Time	GPM Flow
Farm	550	157'	10"	12'	34 sec	1,376
Church	705	375'	10"	8'	28 sec	1,380

Both hydrants were located at elevations of approximately 750 feet above sea level. While both ponds are shallow, their levels remain relatively constant, so strainers were located approximately 1.5 feet below the surface of the water. While not meeting the NFPA's and the ISO's recommendations, having 18 inches of water above the strainer normally enables a pumper to supply its rated capacity without creating a whirlpool. A measuring stick was driven into the bottom of the pond with markings at one-foot intervals to monitor the level of the water near the strainer.

PRIMING THE PUMP

Would 10"-pipe make it difficult to prime the pump? During the tests it took from 35 to 48 seconds to prime the pump, more than the 30 seconds for priming specified by NFPA 1901, but well under the manufacturer's 60-second priming limitation. Pump operators should expect longer priming times for dry hydrants, and not cut off the primer prematurely. On the other hand, if a pump does not prime within 60 seconds, you should check for air leaks or other problems before you continue to operate the primer.

Suction hose tends to sag in the middle when it fills with water and the elbow on the dry hydrant is the highest point. If air is trapped inside the dry hydrant, it can cause the pump to lose its prime when the water starts to move. If you sense that this is happening, operate the primer momentarily, and continue to flow water; this should scavenge the air from the dry hydrant, and allow the pump to deliver a continuous stream of water.

The Bostian Heights department had a brick enclosure built around the dry hydrant near the church to provide additional support as shown in Figure 7-29. A cabinet was built into the brick wall to store adapters and fittings that might be needed for drafting.

Figure 7-29. The masonry structure not only protects the dry hydrant, but has a cabinet for adapters and accessories needed to use the hydrant.

Figure 7-30. Clogged strainer on dry hydrant after years of use.

MAINTENANCE

Dry hydrants tend to accumulate silt or debris, either while they are in use, or when they have not been used for some time. Figure 7-30 is a photograph of a strainer on a dry hydrant. In this case the water level in this pond was lowered for repairs, and the strainer was exposed. You can see that many of the holes are blocked, significantly reducing the maximum flow from the dry hydrant.

Before one of Calvert County's dry hydrants had its fire department fitting installed, an obstruction developed. When the hydrant was installed and flow tests were made, it would supply only 656 GPM. After the hydrant was backflushed with the fire pump, it yielded a flow rate of 1,122 GPM. To be certain that dry hydrants will perform adequately, backflush each on a routine basis. Consider testing and flushing on a quarterly basis. Experience may show that the interval between testing and flushing has to be shortened, or it may be possible to extend it in cases where no deterioration has been shown over a three-month period. Figure 7-32 is an example of a worksheet that can be used to record the test results, and maintenance done. One fire department depends on a dry hydrant which has been installed in a

Figure 7-31. Pumper drafting and applying pressure to dry hydrant to flush it.

DRAFTING HYDRANT TEST RECORD

WATER SUPPLY POINT_____   LIFT_____

LOCATION_____

GROUND ELEVATION_____ DATE OF INSTALLATION_____

DIAGRAM

DRY HYDRANT TEST RECORD						
RECORD OF TESTS		PUMP PRESS.	PRIME TIME	SIZE TIP	NOZZLE PRESSURE	GPM FLOW
DATE	TESTER					

MAINTENANCE RECORDS:

Figure 7-32. Dry hydrant test record worksheet.

shallow river, with the strainer buried in a bed of gravel in the river bed as illustrated in Figure 7-33. In order to minimize the silting effect from the accumulation of sand and dirt in the gravel, it has established a routine of backflushing this hydrant by connecting to the 4½″-drafting outlet with a 2½″-line and pumping 2,000 gallons of water from a tanker through the system at least once a month. Figure 7-31 shows a

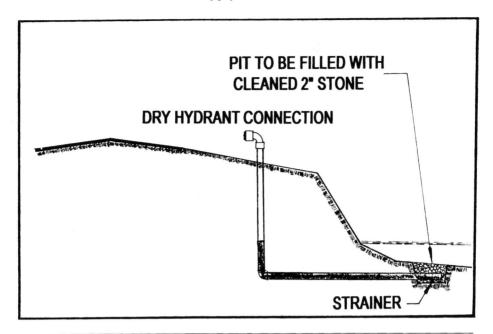

Figure 7-33. Drafting hydrant installed in a stream bed.

dry hydrant being backflushed by using the pumper to draft from the pond, and flowing water under pressure back through the drafting hydrant.

USE OF A PORTABLE DRY HYDRANT IN FREEZING WEATHER

In some parts of the country, a thick layer of ice can freeze on the surface of water, preventing fire department access to draft. To overcome this difficulty, some fire departments have had to depend on breaking a hole through the ice with an axe while others have used a chain saw for this purpose. One solution has been to anchor a plastic bucket partially filled with anti-freeze so that it will float on the surface in an area that can be reached by the suction hose from the pumper. A pike pole then can be used to break through the bottom of the bucket and gain access to the water for drafting.

One fire department that operates in an extremely cold climate uses a portable dry hydrant for access through the ice. The portable dry hydrant consists of a standard dry hydrant connected to approximately 4 feet of 6''-PVC pipe. In the bottom two feet of pipe are a number of holes to serve as a strainer, and there is an adjustable bracket fastened

Figure 7-34. Drafting through a hole cut in the ice.

Figure 7-35. Portable drafting hydrant with ice auger.

around the pipe to determine how far the strainer extends below the layer of ice. This bracket can be adjusted to accommodate the thickness of the ice and the depth of the water at the drafting point. To use the drafting hydrant, a gasoline-powered ice auger is used to drill a hole in the ice. When the portable hydrant is inserted into the hole, the strainer extends below the ice and provides a readily accessible supply of water to the pumper.

TRAINING

Because using a dry hydrant for drafting is a relatively new operation for many fire departments, pump operators need to be trained and to gain experience in the process of drafting through a dry hydrant. This experience for all pump operators, mutual aid as well as first due, is best gained by connecting to and pumping from a dry hydrant on a regular basis. Detailed instructions about the locations of dry hydrants and how to draft from them should be supplied to any departments that may have occasion to use them.

PUMP USE AT DRAFTING SOURCES

Large pipe connected to a dry hydrant allows a pumper to draft from a remote source, but does not help if the lift from the surface of the water to the drafting inlet on the pumper exceeds 20 feet, a problem shown in Figure 7-6. When lift approaches 20 feet, two-thirds of the atmospheric pressure is used to overcome the head pressure, leaving only two PSI available to compensate for friction loss in the system, and severely limiting the pump's maximum capacity. Using a pump at the source to develop adequate pressure to overcome the lift and friction loss in the line, rather than depending on atmospheric pressure, is one solution.

PORTABLE PUMPS

Rural fire departments frequently use portable pumps to draft from ponds or streams that are not accessible by motorized fire apparatus. However, the characteristics that make a pump portable—small capacity and lightweight—also mean that a typical portable pump can supply a maximum of only 250 GPM, not enough to support an aggressive

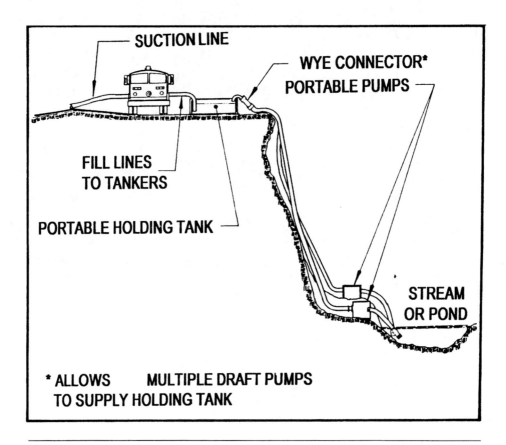

Figure 7-36. Multiple portable pumps and portable tank for water supply.

attack in most cases. The 250-GPM supply also limits the fill rate when loading tankers.

A somewhat successful solution is to use a portable tank for a reservoir and multiple portable pumps to fill it (see Figure 7-36). If you use a pumper to draft from the tank, you can provide a peak flow for fire attack, or for filling tankers, that approaches the rated capacity of the pump. However, among the problems associated with this method of operation are that the maximum flow is limited by the size and length of the supply lines between the portable pumps and the supply tank, and multiple supply lines are needed to transport enough water to supply a pumper.

A standpipe installed at critical locations can eliminate this problem (see Figure 7-37). By installing a dry hydrant, with the supply end terminating in a standpipe connection instead of extending into a strainer submerged under water, portable pumps can be used to supply enough pressure to overcome the lift and friction loss in the standpipe.

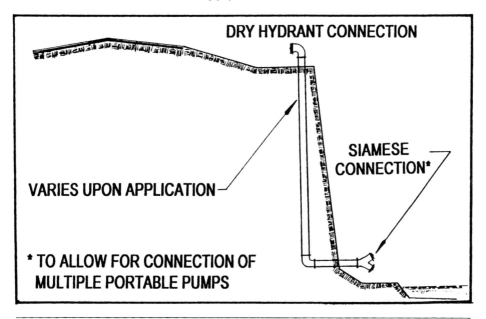

Figure 7-37. Standpipe supplied by portable pumps to overcome excessive lift.

This approach provides a dependable water supply where a pumper would not be able to draft because of the lift. A variation on this arrangement is to install a permanently mounted fire pump at the source using either an electrically driven pump or some other power source for driving it as needed.

HIGH-CAPACITY PORTABLE PUMPS

Among the readily available portable pumps with more than 250-GPM capacity, one of the most useful is a floating pump that can supply more than 400 GPM. Overall, these pumps are simple to operate. They supply their rated capacity in under six inches of water if the recharge or flow rate is sufficient to provide a dependable supply. If you operate two of these pumps in parallel, you can supply more than 800 GPM to a pumper for filling tankers or applying directly to a fire.

There are other portable pumps available with capacities of as much as 500 GPM. Because these units are heavy, and require three or four people to carry and put them into service, you can use trailer-mounted units that you can tow to a water source with a tractor or four-wheel drive unit. Figure 7-38 shows a trailer-mounted pump that is used in a semi-permanent installation with an overhead fill line for loading tankers.

Figure 7-38. Trailer-mounted portable pump with fill device for tankers.

SUPPLEMENTAL SUCTION SUPPLY

You need suction supply primarily in areas where there are no pressurized fire hydrants. Suction supply also can be used to supplement a pressurized hydrant system if the available flow cannot supply the needed fire flow. In ISO evaluations, suction supply is recognized as a means of supplying needed fire flow when available flow from fire hydrants is insufficient.

Many public water systems are not able to supply the needed fire flow in all areas. This happens frequently when commercial or industrial buildings are constructed and served by older distribution systems. It is preferable to install new water lines large enough to meet the community's needs, but this is not always practical. If ponds, streams, lakes, or other sources are readily available, they can supplement the water system's flow, generally at a much lower cost than upgrading the public system. In extreme cases, holding facilities such as dams, storage tanks or cisterns, and dry or wet hydrant systems can be constructed to meet the needed fire flow.

Suction Supply as a Water Source

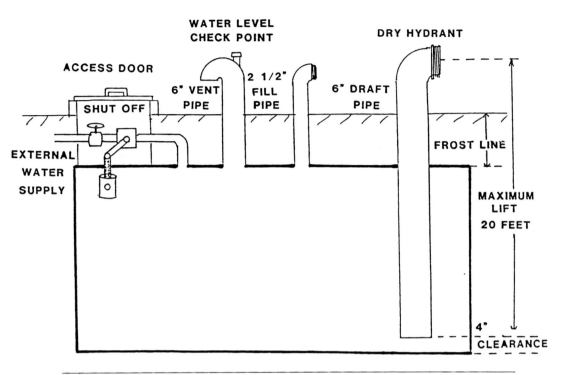

Figure 7-39. Installation plan for underground storage tank with external supply from a domestic system.

Figure 7-39 is an example of an installation plan for an underground storage tank. The tank in this example was designed for an area whose central water system used 4″-lines that were as much as 5 miles long. The distribution system could not supply enough water for a pressurized fire hydrant, but no alternative sources were readily available. Additionally, water was supplied by a special-purpose water district whose board of directors was reluctant to furnish fire protection water at no cost. The tank was installed to satisfy both of these concerns.

A standard dry hydrant allows a pumper to draft from the tank, and a 6″-vent was included to ensure that enough air could be drawn into the tank to prevent a vacuum from developing when water is withdrawn at a high rate of flow. The opening in the pipe from the dry hydrant is far enough above the bottom of the tank to prevent any debris or foreign objects from being drawn into the pump. The tank normally is filled through a standard water service using a meter to measure the amount of water used, a float to shut off the water when the tank is full, and a manual valve. If there is no public water system, a well can be used for refilling. Because the flow from the external supply is limited, it can take several hours to replenish the supply after a tanker is loaded

Figure 7-40. Access to underground storage tank with 6''-pumper drafting connection and vent.

Figure 7-41. Access to underground storage tank with 5''-pumper drafting connection.

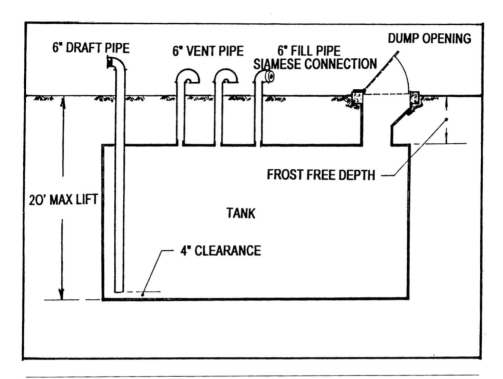

Figure 7-42. Installation plan for underground storage tank with no external supply.

or water is used to supply attack lines. A supplemental fill connection, equipped with 2½″-NST fittings, allows tankers to supplement their supply from the public water system for extended use periods.

If no water source is available, you can use an independent storage tank. Figure 7-42 illustrates a storage tank designed to be used as a reservoir for water supplied by tankers. Because all water is provided by tankers, the 2½″-fill connection uses a clappered siamese, similar to the standpipe fittings frequently found on large buildings. The clappered siamese allows two tankers to unload at the same time, or a water supply tanker to discharge through dual lines. This installation was designed for an area where most of the tankers use large gravity dump outlets for unloading. To accommodate them, a dump inlet was designed with access doors that serve as deflectors to direct the discharge into the dump opening in the tank when they are opened. Two 6″-vent pipes accommodate the high dump rates that are possible using water supply tankers for filling the tank.

This installation was designed to provide protection for a commercial building in a rural area. The dry hydrant was located where it could be used by an engine company to make an initial attack on a fire

Figure 7-43. Above, dry hydrant provides protection for commercial building.
Figure 7-44. Below left, Water Supply Point warning sign.
Figure 7-45. Right, Water Supply Point directional sign.

Suction Supply as a Water Source 211

in the structure. With 10,000 gallons of water readily available, the attack can begin as soon as the first engine company arrives. As tankers arrive, and a water shuttle is established, the underground tank serves as a reservoir, eliminating the need for the fire department to set up a portable tank or use one of the available tankers as a nurse for storage at the dump site.

WATER SUPPLY POINTS

Using suction supply sources, either as a primary or supplemental source, depends on advance planning. Advance planning begins with an analysis of potential water sources and the assigning of priorities. You must get the property owner's permission to use the source and get access for fire department apparatus. You also need signs that warn against blocking fire department access to the water supply point during an emergency, as shown in Figure 7-44, and to direct mutual aid apparatus to a specific location. Regular tests to verify that the supply is adequate should be conducted, and any dry hydrants should be flushed regularly and tested for maximum flow. Pump operators and fire officers who might have to use the water supply point need detailed information and a comprehensive training program.

Suction supply sources can be just as important as fire hydrants, and deserve just as much attention and emphasis as the public water system in prefire planning.

Figure 7-46. Community reservoir for fire protection.

8. Relay Operations

OVERVIEW

You can use a relay for water supply any time your water source is more than a few hundred feet from an incident scene. A relay can be as simple as a two-piece engine company, i.e., an attack pumper on the fire scene supplied by a pumper located at the source. It might consist of setting up a dump site for a water shuttle, and using a water supply pumper to relay water to the attack pumper. At the other extreme, it might be necessary to transport water a mile or more by using multiple pumpers and multiple hoselines. Regardless of the complexity of a relay, the principle is the same: fire pumps are connected into supply lines at intervals to compensate for pressure losses that occur as water moves through the fire hose from the water source to the fire.

PRESSURE LOSSES

Pressure losses in a relay can be static or dynamic.

Static pressure losses are caused by gravity. Water weighs 8.35 pounds per gallon. When an attack pumper is at a higher elevation than its supply pumper, the water's weight in the hoseline creates a head pressure (HP) that must be overcome to supply water to the next pumper in line. The head pressure, or the elevation factor as it is sometimes called in relay calculations, is measured in pounds per square inch (PSI), and is proportional to the difference in elevation between adjacent pumpers. A column of water one square inch in area, and one foot high, weighs .434 pounds. A pressure of .434 PSI is needed to overcome the head pressure, and raise a column of water one foot. For practical fireground hydraulics calculations and relay design, you can round off the anticipated pressure loss to .5 PSI per foot, with an allowance of 5 PSI pressure loss for every 10 feet of elevation. Use this factor only to estimate pressure differentials created by the weight of water and differences in elevation. The volume of water you are transporting has no bearing on the amount of head pressure in the relay.

Friction causes **dynamic** pressure losses which are proportional to the velocity of the water as it flows through the hoselines. As water moves through the hose, portions of the stream that touch the hose lining are slowed down, while water flows freely through the center of the waterway. As the drops of water slide along the hose lining, they tend to roll rather than flow, causing turbulence as illustrated in Figure 7-7 in Chapter 7: Suction Supply as a Water Source. This in turn causes more opposition to the flow, requiring more pressure to overcome it. The term friction loss (FL) is used because the reduction in pressure is caused by the friction between the water and the hose lining as the water moves through it. The higher the velocity of the water, the more friction loss it encounters.

Friction loss in a given layout is affected by the diameter of the supply line, the length of hose used, and the amount of water flowing through it, as well as the appliances being used, gaskets protruding into the waterway, kinks in the line, or any other irregularities. In a given section of hose, velocity, turbulence, and the resulting friction loss all increase as the amount of water flowing increases. The friction loss increase is proportional to the square of the amount of increase in flow. For example, if you double the flow, four times as much pressure loss will be generated. A 200-GPM flow generates approximately four PSI per 100' in 3"-hose; increasing the flow to 400 GPM creates a friction loss of 16 PSI. The rapid increase in friction loss with high rates of flow limits the maximum GPM that can be transported through any given hose layout. See Chapter 4: The Use of Fire Hose in a Water Delivery System, of this book for information on hose characteristics and practical methods of calculating anticipated friction loss.

Calculating Pressure Requirements

At the terminal end of the relay, determine the required discharge pressure for the attack pumper by looking at the pressure requirements for the attack lines and appliances that will be used. Determine the optimum pressure for each of the preconnected attack lines under controlled conditions. Use company drills or training sessions to experiment with various pressures, nozzle settings, and hose layouts.

Set the engine pressure on the attack pumper for the highest pressure needed to supply any of the preconnected attack lines. You can partially close discharge valves and lock them in position to increase the friction loss; this allows you to reach the desired operating pressure for any of the attack lines that require a lower operating pressure.

When additional attack lines are needed, you can determine the required engine pressure by adding the required nozzle pressure to the anticipated friction loss in the hose and appliances. Then, calculate using

$$Ep = FL \pm HP + NP$$

where

- Ep = the discharge pressure at the pump required to attain the desired pressure at the nozzle
- FL = the total friction loss between the pump and the flow device
- HP = the head pressure generated by the difference in elevation between the fire pump and the location where the flow device will be used. (One way to estimate head pressure is to allow 5 PSI for each story of the fire building, including the ground floor and the fire floor)
- NP = the required operating pressure for the flow device being used. Operating pressure normally is specified by the manufacturer, but can be determined more accurately by experimenting with actual flow under typical operating conditions

For the supply pumper, and any relay pumpers, you must first calculate what pressure losses need to be overcome before you can calculate the required engine pressure. Once the variables are known, you can determine the desired discharge pressure by

$$Ep = FL \pm HP + 50$$

where

- Ep = the engine pressure needed to overcome the losses in the hoselines while maintaining a safe residual pressure at the next pump in line
- FL = the total friction loss to be expected in the line between pumpers. (This can be determined by any of the methods described in Chapter 4 of this book)

HP = the head pressure, based on an estimation of the difference in elevation between adjacent pumpers in the relay. (For practical purposes, the engine pressure should be set to allow for a 5-PSI differential for each 10 feet of elevation)

50 = the 50-PSI residual pressure at the intake of each relay pumper and the attack engine that the formula provides for. Designing a relay for a residual pressure of 50 PSI at the intake of each pumper provides a margin for error that will compensate for any gauge inaccuracies, and will accommodate small changes in flow without causing the supply line to collapse

RELAY ORGANIZATION

Successful relays must be organized. After a relay has been designed, the hose is on the ground, the pumpers have been committed and the water is moving, it is too late to make any changes. The incident commander is too busy to plan and establish a complex relay. That job is up to the water supply officer (WSO) who should be designated as soon as the officer in charge arrives. From that point on, the incident commander remains isolated from the water supply operation. Tell the WSO what is required, and give him or her the freedom to do whatever is necessary to provide enough water to supply the needed fire flow.

Water Supply Officer

The WSO needs full authority to set up the relay, including assigning the necessary apparatus, determining hose layouts, arranging for mutual aid when required, and exercising operational control of the relay. When a complex water supply problem requires more than one relay, supplemental supplies, or both, the WSO may have to appoint a separate relay control officer (RCO) to deal with each one. If so, the most important requirement is that a clearly designated authority and chain of command coordinate each operation.

Relay Control Officer

Each unit in the relay must know who is in charge, where the RCO is located, and the best means of communication. The RCO makes all decisions and issues all commands and directions.

The RCO also must be in charge of the relay's water usage; no additional attack lines should be put into service without the RCO's knowledge and concurrence. By knowing the relay's maximum capacity, and how much water is flowing, the RCO can determine whether supplying additional lines is possible. By knowing what has to be done to obtain additional water from the relay and how long it will take to deliver it, the RCO is positioned to help the WSO and incident commander set use priorities for available water.

DESIGNING A RELAY

There are certain limitations inherent in any relay operation. Pumpers are designed to supply their rated capacity at a maximum net pump pressure of 150 PSI. Higher discharge pressures reduce the capacity of the pump accordingly. What also limits pressure in a relay is the maximum working pressure of fire hose. Most fire hose is tested annually at a pressure of 250 PSI. However, given that a 50-PSI safety factor is needed to allow for deterioration of the hose and minor pressure variations during operation, maximum working pressure in a relay should be limited to 200 PSI. If higher pressure is needed, pumpers can develop as much as 250 PSI. The question is whether the hose can withstand it. When operating at a net pump pressure of 250 PSI, however, the pump operator needs to know that the maximum flow capability of the pumper will be reduced to 50% of its rated capacity. As was explained in Chapter 5 and noted in Table 5.1, a pumper is designed to supply 50 percent of its rated capacity at a pressure of 250 PSI.

Most fire hose is manufactured to meet a specified test pressure of 600 PSI, but deteriorates with use and age. If your department frequently needs working pressures higher than 200 PSI, your annual hose tests should be at 300 PSI. The higher test pressure gives you a working pressure for relays of 250 PSI, yet retains the 50-PSI safety factor. If your manufacturer has specified a lower test pressure, do not exceed it during annual tests. Large Diameter Hose (LDH) frequently is manufactured with a pressure test limit under 300 PSI; the maximum relay pressure is reduced accordingly. Always allow for the safety factor by keeping the maximum working pressure in a relay at least 50 PSI below the hose's test pressure.

All friction loss calculations are based on the amount of flow through the relay. If adjustable or automatic nozzles are being used, the actual flow through a given attack line is controlled at the nozzle, not at the pump. Without pumper flow meters, the water supply officer cannot determine how much water is actually flowing at any given time. With the flow controlled at the nozzle, your relay must be designed to supply a flow that is based on the nozzle being set for maximum discharge, not what is actually flowing. A pump operator cannot practicably change the pressure at the fire pump each time a nozzle setting is changed or a line is shut down during the fire attack, so if more water is needed, it must be available. Table 4.5 (Chapter 4) provides estimated maximum flows for some typical fire attack lines. However, for relay design purposes, expect the maximum flow rates in Table 8.1.

Table 8.1. Approximate Flow Rates for Attack Lines

Size of Attack Line	Maximum Flow Rate
Initial Attack and Backup Handlines	
1½"-Line	120 GPM
1¾"-Line	150 GPM
2"-Line	200 GPM
2½"-Line	250 GPM
Master Stream Supply Lines for Heavy Attack	
2½"-Line	375 GPM
3"-Line	550 GPM
Large Diameter Hose	1,000 GPM

An automatic nozzle may exceed these estimated flow rates when a higher-than-normal pressure or very short hoselines are used. The rates do, however, provide a good basis for designing a relay. By using standard flow ratings for different size attack lines, the relay design is based on the hoselines connected to the attack pumper rather than on the amount of water flowing. This simplifies pressure calculations, and ensures that the relay will be able to meet any demands the fire attack may impose. If additional attack lines are needed, the relay will have to be modified to supply them.

Determining the Pumpers and Hoselines to be Used

You design your relay based on needed fire flow, available pumpers, and type of hose. With the needed fire flow known, source of water selected, and the distance to the incident scene estimated, you are ready to calculate total pressure losses. For example:

Needed Fire Flow	1,000 GPM
Distance from Supply Point to Fire	2,000 feet
Difference in Ground Elevation	20 feet
Capacity of Available Pumpers	1,000 GPM
Unlimited supply of 2½"-hose	

(Note: The hand "method" and the conversion factors listed in Chapter 4 have been used for all friction loss calculations.)

A 1,000-GPM flow through a 2½"-fire hose could be expected to generate approximately 200 PSI per 100 feet, so it would be impossible to use a single 2½"-line for this relay. You need parallel lines so the flow can be split between them with each line carrying 500 GPM. This would reduce the friction loss to 50 PSI per 100-foot section. To stay within the 150 PSI net pump pressure limitations, you need to insert a relay pumper in line every 300 feet. This operation would require a total of at least eight pumpers. Adding a third parallel line would bring the flow in each line down to 333 GPM, and reduce the friction loss to approximately 20 PSI per 100 feet. With the reduced friction loss, you could space the pumpers 700 feet apart, and need only five instead of eight. If a fourth line were added, each line's flow would be 250 GPM, friction loss would be reduced to 12 PSI in each 100-foot section, and only three pumpers would be needed to supply the needed fire flow. With a 20-foot difference in ground elevation over the entire relay, the total loss from head pressure would be 10 PSI, and the relay pumpers in each of these examples would be able to handle the additional needed pressure. These examples show that it takes many pumpers to move more than 250 GPM any distance through a single line of 2½"-hose.

Pumpers have the disadvantages of being expensive, requiring a driver, and not always being available, and, in relay operations, each pumper in the line is a potential problem that can adversely affect reliability and efficiency. On the other hand, laying four parallel lines to move 1,000 GPM is time-consuming and inefficient. The most practical solution? Use larger hose for supply line.

In the same example, if you used 3"-hose equipped with 2½"-couplings instead of 2½"-hose, you could reduce the number of parallel lines. In a single line of 3"-hose, friction loss would be 100 PSI per 100 feet, still an impractical layout. Dual lines would reduce the flow in each line to 500 GPM and the friction loss per 100 feet to 25. This way pumpers could be spaced 600 feet apart, andfive pumpers would be adequate to handle the flow. Adding a third line would bring the flow down to 333 GPM in each line, and the friction loss would be approximately 10 PSI per 100 feet. In this case you would need only one relay pumper between the supply pumper and the attack pumper, but you would still need multiple lines.

Large diameter hose is the only way to eliminate the need for multiple supply lines when you need a fire flow greater than 500 GPM more than 500 feet from the source. If a supply of 5″-LDH is available, the problem posed in this example is easier to solve. A thousand gallons per minute flowing through 5″-LDH has an expected friction loss of approximately six PSI per 100 feet. With a total friction loss of 120 PSI, and a head pressure differential of 10 PSI, the supply pumper could easily supply the needed pressure to the attack pumper without a relay pumper in the line. Two pumpers carrying 2,000 feet of 5″-LDH could supply as much water as 8,000 feet of 2½″-hose and three pumpers for this distance.

We tend to assume that friction loss in large diameter hose is negligible, that it does not need to be considered in establishing a water supply. This is true only when small volumes of water are required. When we encounter large fires and must use long hoselines, friction loss still can be a limiting factor. It is especially important to include friction loss in the water supply design when 4″- instead of 5″-LDH is used.

In our example, a single line of 4″-LDH carrying 1,000 GPM would have 20 PSI of friction loss in each 100-foot section. You would have to space pumpers in the relay every 700 feet, and you would need four pumpers to supply 1,000 GPM, but only two if 5″-LDH were used. With dual lines of 4″-LDH, friction loss would drop to five PSI per 100 feet, and no relay pumper would be needed, but, again, multiple lines would be required.

When you need more water, there is a tendency to increase the pressure in a relay rather than lay additional supply lines. This not only limits the amount of increase in flow, but reduces the capacity of the pumper to handle additional lines when the discharge pressure is higher than 150 PSI net pump pressure. Adding a second 1½″-line to a pumper already supplying one 2½″- and one 1½″-line from a supply line consisting of 1,000′ of 3″-hose (an increase of only 25%) would force the pressure required at the supply pumper to climb by almost 50%, from 175 PSI to 262 PSI, **more than a pumper is rated to supply**. This would exceed the safe working pressure in the relay.

Pumping Requirements

When you design a relay, you assume that each pumper can develop 150 PSI net pump pressure at its rated capacity. Net pump pressure is the difference between the intake and discharge pressure and is a measure of the amount of pressure being developed in the fire pump. To deliver a residual pressure of 50 PSI to the intake of the next relay pumper with 150 PSI of pressure losses between them, a pumper would have to maintain a discharge pressure of 200 PSI, which, with an intake pressure of 50 PSI, only requires a relay pumper to develop a net pump pressure of 150 PSI. The pressure chart (Figure 5-4) in Chapter 5: Testing Water Supply Apparatus, provides a means of calculating net pump pressure. By converting PSIg readings to PSIa, and subtracting the lowest number from the highest number, you can determine net pump pressure. While the attack engine also can take advantage of the residual pressure to reduce the net pressure the pump is required to develop, this may not be true of the pumper supplying the relay. If the supply pumper is operating from a hydrant, it can take advantage of the residual pressure in the water distribution system, and the net pump pressure will be lower than the reading on the discharge gauge. However, if the supply pumper is operating from draft, this is not the case, because not only will the pumper have to develop the full discharge pressure, it also will have to provide the needed negative pressure, measured in inches of vacuum, to overcome the losses incurred getting the water from the source into the pump. As a rule, put your largest pumper at the source to provide enough capacity. In this example, if the supply pumper is operating from draft with a 10- foot lift, the required net pump pressure would be approximately 210 PSI. Therefore, a 1,500-GPM pumper would have the extra capacity needed to supply 1,000 GPM to this relay at the higher net pump pressure.

When you design a conventional closed relay, remember that every transported drop of water moves through the entire relay. Your maximum available flow is limited by the smallest pumper's capacity in the relay and by the flow capabilities of the hoseline in the most restrictive section. When you need to use a lower capacity pump or smaller hoselines in one section of the relay, you can shorten the distance between pumpers to allow for the lower pump capacity or greater friction loss in the smaller hoselines.

Engine Pressure Calculations

After you establish the relay, each pump operator needs to determine the required discharge pressure. In a well-organized supply operation, the RCO or WSO specifies each pumper's operating pressure. In other cases, each pump operator calculates the discharge pressure individually by estimating the distance to the next pumper in line and the differences in ground elevation. If a design flow for the relay has been specified, use that figure to calculate friction loss. Without a specified required flow, use a 250-GPM flow for 2½"-hose, 400 GPM for 3"-hose, and 800 GPM for 4"-hose. For 5"-line, use the maximum rated capacity of the pumpers in the relay as a basis for these calculations.

Pump operators need to be familiar with the amount of hose a pumper normally carries to estimate the length of hoseline to the next pumper. If you can estimate the amount of hose still left in the hose bed, you can approximate the number of feet already removed.

In the example of pressure calculations in Figure 8-1, an 800-GPM flow is needed at a point 2,000 feet from a pond located approximately 20 feet below the attack pumper and 10 feet below the relay pumper. A water supply pumper is drafting from the pond, supplying a single line

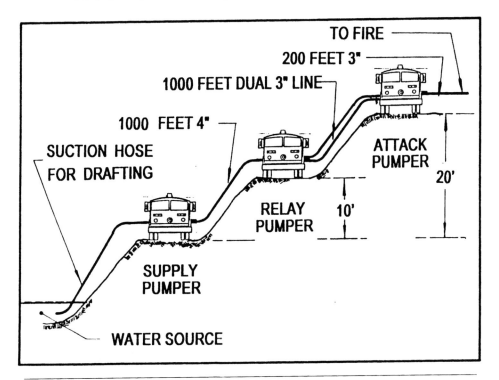

Figure 8-1. Relay problem moving 800 GPM 2,000 feet.

of 4″-hose to a pumper 1,000 feet away. The relay pumper is supplying the attack pumper through a dual line of 3″-hose, and the attack pumper is supplying a master stream device through a dual line of 3″-hose 200 feet long.

SUPPLY PUMPER CALCULATIONS

EP	=	FL	±	HP	+	50
EP	=	130	+	5	+	50
EP	=	185				

RELAY PUMPER CALCULATIONS

EP	=	FL	±	HP	+	50
EP	=	160	+	5	+	50
EP	=	215				

ATTACK PUMPER CALCULATIONS

EP	=	FL	±	HP	+	NP
EP	=	42	+	0	+	100
EP	=	142				

NOTE: 1. The friction loss calculations were made by using the hand method as specified in Chapter 4.
2. Nozzle pressure on the master stream device is assumed to be 100 PSI.
3. A friction loss of 10 PSI was included in the calculations to allow for loss in the inlet to the master stream device.

Operating Practices

First, connect all hoselines to the appropriate discharge and intake connections. Do not connect hose larger than 2½″ to the gated auxiliary suction inlet because these inlets generally are installed using 2½″-pipe with elbows and other fittings in the line. For flows greater than 250 GPM, you may encounter excessive friction loss in the gated intake to the pump, which will limit the maximum flow through the relay. Using a suction siamese, a reducer, or other adapter on the large suction connection is more efficient and reliable under changing flow conditions. If you are using more than one supply line, connect each to a gated fitting, or an external gate valve. You must be able to individually control supply lines in case a line breaks or is taken out of service for some reason. In an emergency, you can shut off the defective line, and maintain a reduced flow with the remaining supply line.

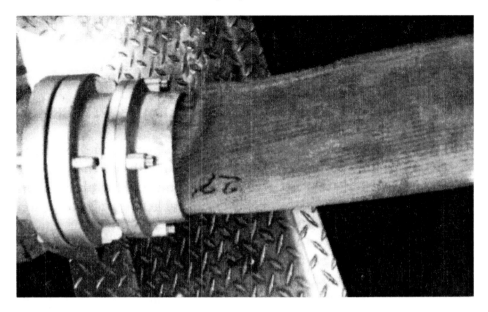

Figure 8-2. Incoming supply collapsed due to excessive flow.

Whenever possible, isolate the intake and discharge lines by connecting them to opposite sides of the pumper. With a standard side-mounted pump panel, a recommended practice is to connect the intake lines to the pump near the pump operator's position so they can be observed closely. Experienced pump operators often maintain physical contact with the incoming supply line. If the flow changes, and the residual pressure drops below zero, the line will first fluctuate, then collapse. A fluctuating or vibrating supply line may be the first symptom of problems in the relay. These physical changes tell the pump operator that the flow of water may be interrupted unless he or she takes some prompt action.

Next, uncap at least one outlet on each of the pumpers to give the pump operator a "safety valve" in case pressure increases dangerously in the relay. While normal pressure variations should be handled by the automatic pressure control device, occasionally operating errors in a previous section of the relay or sudden changes in flow can cause a pressure surge. If you shut down all attack lines simultaneously, and the attack pumper does not maintain the flow by dumping water, friction loss in all sections of your relay will drop to zero. While the automatic pressure control devices on each pumper limit the pressure increase, each pump adds a minimal amount to the residual pressure it receives. If the supply pumper is operating at a pressure of 200 PSI or more, pressure at the relay pumpers could reach a dangerous level quickly. If you open the "safety valve," water flows through the relay to that point, where it is discharged. Friction loss in the supply line reduces the residual pressure at the pump to a safe level.

CAUTION: OPENING THIS VALVE WHILE THE RELAY IS OPERATING IS AN EMERGENCY MEASURE ONLY! IT WILL AFFECT THE PRESSURE THROUGHOUT THE RELAY AND NORMAL OPERATION WILL BE DISRUPTED.

It is important to note that this outlet may discharge large quantities of water under high pressure without warning, and the discharge point should be under the direct observation and control of the pump operator at all times. In some cases, it may be desirable to connect a section of 2½"-hose to use as a dump line; that way you can direct the flow of water to a safe area. If you use a dump line, secure the end of the line to a solid object; this prevents unwanted motion when water is discharged suddenly. Position the dump line so that water being discharged drains properly and does not injure personnel or damage surrounding objects.

You can use the dump line, or uncapped discharge, to build the relay one section at a time, especially important when the supply pumper is operating from draft. By opening an uncapped discharge outlet, and allowing water to flow back into the water source, you can prime the pump, build up pressure, and set the automatic pressure control device while you establish the rest of the relay. When the first pumper in the relay is ready for water, the supply pumper will be ready to provide it. This eliminates delays caused by difficulties in priming the pump and establishing the draft. You can use the same procedures at relay pumpers. Let each flow water as soon as the supply lines are in place; this allows the water to escape through the dump line until the next pumper in line is ready to receive it.

When all the lines are in place, and uncapped discharge or dump lines are open on all relay pumpers, charge the pump at the source and set the required operating pressure. If the supply pumper is operating from draft, prime the pump and discharge water into the source until the first relay pumper is ready for it. After establishing the required engine pressure on the source pumper, and when the next pumper in line is ready, open the discharge valves. Do this slowly so the water can begin to move without causing dangerous pressure surges or water hammer. Another reason to open the line slowly is that empty hoselines present virtually no resistance to the flow of water. This means that if the pumper is operating from draft, there is a good chance the pump may lose its prime if the valves are opened suddenly. With a hydrant, the sudden change can cause damaging water hammer. As you open the valve that controls the outlet to the discharge line, close the safety valve that controls the dump outlet in a coordinated movement. If necessary, adjust the hand throttle on the pumper to maintain the desired operating pressure as the valve is opened and the flow through the pump increases.

When the supply line is charged, the air forced out of the empty hoselines as they fill with water is exhausted through the open dump outlet at the next pumper in line. When the air is gone and a steady stream of water flows from the dump outlet, it is time for the relay pumper operator to engage the pump and adjust the throttle to attain the desired operating pressure. At the same time, he or she should adjust the safety valve to keep the incoming pressure from dropping below zero. When the next pumper in line is ready for water, open the discharge valves to the supply line slowly and close the valve to the dump outlet in a coordinated effort. Never let the compound gauge on the intake drop below zero. A vacuum reading on the intake indicates that too much water is being discharged, and the supply line will collapse (Figure 8-2). If the compound gauge does drop below zero, partially close the valves on the discharge until the pressure stabilizes and begins to rise. If you open the discharge valve too quickly, the pump will try to fill the entire line at once; this requires more water than the relay generally can supply. Attempting to supply water to the supply lines while the dump outlet is still open is another common problem. The dump outlet is used only to get rid of water when there is excessive pressure buildup or to maintain the design flow in the relay until the next pumper is ready to receive it. If the relay has been established and water is moving, you should never need to open the dump outlet unless there is excessive pressure in the relay.

Each relay pumper in turn follows the same procedure until all sections of the relay are in place and water reaches the attack pumper.

You can use bleeder valves, located on the line side of intake valves, to allow air to escape from the line as it fills with water; this prevents it from being forced into the fire pump. Leave the bleeder valves open until all air has been exhausted, and you can see a steady stream of water coming from the bleeder opening. If you are using the fire pump to supply attack lines from the water tank on the pumper while waiting for water from the relay, air from the supply lines, passing through the pump, could cause a pressure fluctuation to the attack lines. In extreme cases, this can cause the pump to lose its prime, interrupting the flow completely. After all air has escaped from the supply line, open the intake valve slowly, while closely observing the discharge pressure on the pump. Since the transition to an external supply changes the pressure on the intake side of the pump, the discharge pressure also may change. When this happens, you must adjust the throttle to compensate for the change in intake pressure to maintain the required operating pressure for the attack lines.

Once proper operating pressure is established throughout the relay and water is flowing, set the automatic pressure control devices (relief valve or pressure governor) on all pumpers in the relay. If any pumpers in the relay are equipped with intake relief valves, they will open and water will discharge when the incoming pressure exceeds the valve's operating point. Water that discharges from a relay pumper can have an adverse effect on the maximum flow through the relay; this causes excessive pressure variations.

A properly adjusted intake relief valve does not open intermittently at normal operating pressure, but rather acts as a safety valve any time excessive pressure is experienced; it allows water to escape and prevents damage to the pump. The increased flow limits any pressure surges by increasing the friction loss in the supply line and reducing the residual pressure at the intake of the pump.

If the flow through any of the attack lines is shut off or reduced, the reduced friction loss in the supply line could increase the residual pressure at the intake enough to prevent the attack pumper from maintaining the desired operating pressure. You can use a dump line connected to an outlet on the attack pumper to increase the flow and reduce the residual pressure at the intake to a manageable level. Adjust the valve on the dump outlet slowly while you closely observe the residual pressure on the intake gauge; make sure that it does not go below zero at any time.

Always put the relay in operation from the source to the attack pumper, or the terminal end of the relay. Reverse this process to shut down. The attack pumper operator reduces the pressure by decreasing the throttle to an idle, then closing all discharge lines. Each relay pumper operator follows the same procedure, in turn, until the source pumper operator shuts down last. If it is possible that the relay will need to be reactivated, the operator of the source pumper can maintain its discharge pressure while opening the dump outlet to allow the water to escape after all discharge lines are shut down.

While the relay is in operation, keep pressure adjustments to a minimum. Minor changes in flow will be reflected in pressure changes throughout the relay, but do not worry about maintaining exact pressures at each pumper. What is important is to make sure that no intake pressure drops low enough to cause the hose to collapse or rises too high to allow the pump to maintain the desired discharge pressure. All pump operators in the relay should monitor the intake pressure, and notify the RCO of any potentially dangerous situation. Keep dump outlets ready to bleed off any dangerously high pressure.

COMMON PROBLEMS

Of the two most frequently occurring relay problems, the most important is the amount of time it takes to begin to supply water. A close second is that even after the relay is set up, it may not give you enough water for the size of the fire.

The first few minutes on the scene of an incident are the most important. If a fire is large enough to require a relay, the time factor is generally critical.

It is not uncommon for a relay to take 30 minutes (or more) to set up. This is too long for most fireground operations. After 30 minutes, the available fire flow from the relay, which might have been adequate to handle the situation on arrival, may be inadequate to control the fire.

Relays that cannot provide enough water to control the volume of fire generally have pumpers that are spaced too far apart, or hose that is too small to supply the needed fire flow. Many fire departments, especially rural departments where long relays are common, depend on 2½"-hose for supply lines. Any appreciable distance limits the total fire attack capability to a single 2½"- or two 1½"-attack lines. Most rural fire departments can provide as much fire flow with a water shuttle using fire department tankers as a single 2½"-line relay. A water shuttle requires much less effort, with virtually no delay in making the initial attack. All things considered, a relay using a single line of 2½"-hose generally is an exercise in futility.

Rural fire departments need to acquire as much large diameter hose as possible to move large amounts of water over long distances with a minimum amount of fire apparatus. If only 2½"- or 3"-hose is available for supply lines, the standard practice should be to lay multiple lines any time a relay is needed.

Even if you're using hose that is three inches or larger in diameter, your maximum flow still will be limited if the distance between adjacent pumpers is too great. You will need to reduce the flow to lower the friction loss per section enough to be within the pumpers' capability to supply it.

STANDARD OPERATING PROCEDURES

Standard Operating Procedures (SOPs) can be written to minimize delays in establishing a relay and to provide adequate flow once a relay is operational.

In writing a set of relay SOPs, consider the following basic assumptions and rules:

- Only consider a pumper for relay operations that can develop a net pump pressure of 150 PSI at its rated capacity. The pumper also must have a relief valve or pressure governor that functions quickly and can be set accurately.
- The distance between adjacent pumpers in the relay should not exceed 1,000 feet. The easiest way for a pump operator to know how much hose has been laid is to carry supply hose racked in beds of 1,000 feet on each pumper. If a single pumper will carry more than 1,000 feet of supply line, you can partition the hose bed, or mark each section of hose as it is loaded so that the pump operator can determine how much hose has been taken off by looking at the number on the last section still in the hose bed.
- The supply lines in the relay must match the attack lines being used:
    - One 2½"-supply line will support one 2½"-attack line or two 1½"-attack lines. (250 GPM)
    - Each 3"-supply line will support one 2½"-attack line and one 1½"-attack line or three 1½"-attack lines. (375 GPM)
    - Each 4"-supply line will support three 2½"-attack lines or a master stream. (800 GPM)
    - One 5"-supply line will support multiple master streams and attack lines. (2,000 GPM)
- The supply pumper, and all relay pumpers maintain a standard discharge pressure of 200 PSI.

With these standards, friction loss between pumpers will never exceed 150 PSI. Unless there are unusual differences in elevation, the maximum pressure in a relay never needs to be higher than 200 PSI to maintain a residual pressure of 50 PSI at the next pumper in line. Using a standard discharge pressure on each pumper of 200 PSI should maintain a safe operating pressure for most hoselines. While the discharge pressure exceeds the 150-PSI pressure limitation for 100% capacity of

Figure 8-3. Four-way valve to insert a relay pumper in line without interrupting the flow.

Figure 8-4. Relay pumper using a four-way valve in line.

the fire pump, there generally will be at least 50-PSI residual pressure at the intake of each of the relay pumpers and the net pump pressure will not be more than 150 PSI. Since the supply pumper may not have as much residual pressure on its intake, it is generally a good idea to assign the largest pumper to this role. This is especially important when the relay is being supplied from draft. In order to develop a net pump pressure in excess of 200 PSI to supply the relay from draft, the rated capacity of the supply pumper should be at least 30% higher than the other pumpers in the relay.

A standard operating procedure helps you to avoid the tendency to increase the pressure to provide additional fire flow. With a standard pressure of 200 PSI on the discharge of the supply pumper and each relay pumper, the full capacity of each pumper, and the design flow capabilities of the hoselines, will be available at all times. If you need more flow than your relay can supply, additional hoselines, additional pumpers, or both, will be needed.

The relay SOP also helps the incident commander to readily determine available fire flow. You can determine the number and type of attack lines to use by observing the size and number of supply lines.

CONSTANT FLOW RELAY DESIGN

Using a relay SOP helps to ensure that a relay will always be able to supply enough water for the attack lines, no matter how the variable flow nozzles are set. The problem in a conventional relay occurs when one or more of the attack lines are shut down, or when the nozzles are set to reduce the flow. With the reduced flow, friction loss in the supply line decreases, and the residual pressure at the intake of each relay pumper increases. If the automatic pressure regulating device on each of the pumpers is set properly and functions well, it limits the amount of increase passed on to the next pumper in the relay. Unfortunately, many pressure governors and relief valves either are unreliable or do not operate fast enough to prevent pressure surges. Even when they are new, they are required only to limit any pressure increase to 30 PSI. When multiple pumpers are connected in relay, the pressure increase could reach dangerous levels and damage hoselines or a fire pump. Even a limited increase in residual pressure at the attack pumper can make it difficult to maintain the proper operating pressure to the attack lines.

Controlling the Flow

The only practical way to maintain consistent operating pressures within a relay is to keep the same amount of water moving. You can do this by using a dump line.

If an attack line is shut down, the residual pressure at the intake of the attack pumper will register an increase that is proportional to the decrease in flow and the reduced friction loss in the supply line. By opening a valve to allow water to discharge from a dump line, you can increase the flow to compensate for the attack line which is no longer operating. For example, you shut down an attack line which is flowing 250 GPM, and the residual pressure increases. Open the discharge valve on the dump line to allow 250 GPM to escape. How much should the valve be opened? To set the valve on the dump line properly, observe the compound gauge on the intake of the attack pumper. When the valve has been opened enough to compensate for the line which has been shut down, the flow will go back to normal, and the residual pressure will return to 50 PSI.

When you open the attack line again, the flow will increase and the residual pressure will drop below 50. This is the cue for the pump operator to close the valve on the dump line. This type of operation requires the pump operator to concentrate and act quickly. He or she uses the compound gauge as a water meter, and adjusts the dump valve to maintain the incoming pressure at 50 PSI ± 30 PSI.

By maintaining a constant flow from the attack pumper, friction loss will remain the same throughout the entire relay and no adjustments should be required at the relay pumpers or at the supply pumper. This simplifies operations and prevents random pressure changes.

Getting Rid of the Water

If you use the constant flow method of operation, water will flow from the dump line during much of the relay operation. Where you locate the point of discharge from the dump line is critical for the following reasons:

1. Water could flow unexpectedly under high pressure, and in a large volume, from the dump line at any time, so, as a safety measure, you need to fasten the line securely to prevent uncontrolled movement. To prevent injuries, keep people away from the point of discharge. When the attack pumper is located on a paved surface, connecting a 2½"-hard suction hose to an unused dis-

charge outlet from the pump, and fastening the end of the hose to a wheel, with the water discharging under the vehicle can be a good solution to this problem.

2. Because large amounts of water can flow from the dump line over a period of time and accumulate, the discharge location must have good drainage. Fastening the dump line to a storm sewer grating is a good way to both get rid of the water and stabilize the end of the hose.

You can reduce the velocity of water being discharged from the dump line by connecting the dump line to the intake side of the pump, through the auxiliary suction fitting, instead of a discharge outlet. Connecting the dump line to the discharge means that the same flow is maintained through the pump on the attack pumper, as well as the rest of the relay; this makes control more effective. On the other hand, the dump line on the intake controls the incoming pressure, so the automatic pressure control device on the attack pumper should be able to maintain the correct operating pressure to the attack lines.

CONSTANT FLOW RELAY OPERATIONS

In a constant flow relay, pressures throughout the relay are controlled by the pump operator at the attack pumper or the terminal end of the relay. When there are frequent changes in flow during the initial attack, the pump operator must watch the incoming pressure closely to compensate for any changes by adjusting the discharge valve on the dump line. As in any relay, the fewer adjustments made, the smoother the operation. As a general rule, if you can keep the incoming pressure to between 20 and 80 PSI, the discharge pressure can be maintained at the proper level. The most difficult part of constant flow relay operation is making a manual adjustment to the valve on the dump line each time the flow changes in any of the attack lines.

AUTOMATIC RELAY OPERATIONS

Many pumpers are equipped with an intake relief valve and the 1991 edition of NFPA 1901 requires them on all new pumpers. This relief valve is connected to the intake side of the pump and is designed to protect the fire pump from excessive pressure during relay operations. When the incoming pressure exceeds the operating point of the valve, it opens and allows water to discharge outside of the pump. When this happens, the intake relief valve functions much like the manual dump line, that is, it reduces the incoming pressure by increasing the flow through the supply line. As the flow increases, the friction loss in the incoming hoseline increases, and the residual pressure at the pump will not exceed the operating pressure of the intake relief valve. Set properly, this valve will automatically maintain the same flow through the relay when changes are made in nozzle settings, or nozzles are opened and closed, and pressure readings should remain constant.

INTAKE RELIEF VALVE OPERATIONS

NFPA 1901 specifies that apparatus manufacturers shall set intake relief systems at 125 PSI, unless otherwise directed by the purchaser. With the operating point set at 125 PSI, incoming pressure cannot go high enough to damage the pump, nor will the relief valve open unexpectedly when connected to a hydrant, or during normal relay operations.

A better practice would be to set the relief valve to maintain the expected residual pressure each time it is used in a relay. If the residual pressure at the pumper with all attack lines flowing is 50 PSI, then the intake relief valve should be set to 50 PSI. As long as all lines are flowing, the valve remains closed. If an attack line flowing 250 GPM is shut down, the incoming pressure will rise according to the reduced friction loss with the lower rate of flow. As the pressure rises, the intake relief valve will open far enough to allow 250 GPM to dump on the ground, maintaining a residual pressure of 50 PSI. If the line is put back into service, the increased flow will cause the residual pressure to go below 50 PSI, and the valve will close. The ideal setting for an intake relief valve would be 50 PSI in this type of operation, a setting too low for most relief valves. NFPA 1901 (1991) requires that the intake relief valve be adjustable from 75 PSI to 250 PSI; some can be set even lower. Although, in some cases, this is something more than the ideal pressure, the intake relief valve can still be useful. It should be able to maintain a workable residual pressure in an operating relay under changing conditions.

Without specific instructions to the manufacturer, the adjustment that controls the operating pressure of the intake relief valve will not be field adjustable from the pump panel. If the controls are not readily accessible, the valve generally will be set to a predetermined operating pressure and left there. If the operating point is set too low, the valve could open, wasting water when the pumper is connected to a hydrant with higher than average static pressure. When the intake relief valve adjustment is not accessible to the pump operator, set the valve to a pressure higher than the highest pressure you might encounter when connecting to a hydrant.

The intake relief valve outlet should be equipped with 2½"-NST fire threads. All new apparatus that meets NFPA 1901 will have a surplus water discharge from the intake relief system equipped with a male fitting with NST threads. Many pumpers in service today do not have a discharge connection accessible to the pump operator. In a relay where the valve can be expected to operate intermittently, the standard connection on the outlet of the valve can be used to connect a dump line to direct the water to an area where it will not cause any harm. It also enables the pump operator to close off the relief valve outlet to prevent unwanted water flow if there is no shutoff valve. Another plus is that you can stop any air from leaking back through the valve if you are attempting to draw a vacuum and operate at draft.

By using the intake relief valve to maintain a constant pressure at the intake of the terminal pumper, and a relief valve or pressure governor to control the discharge pressure on each of the pumpers in the relay, operation becomes automatic. With all pumpers properly set up, it should be possible to vary from maximum flow to no water discharging from the relay without manually adjusting any of the pumpers.

Limitations

A limited water supply is the only condition under which the constant flow principle and automatic relay concepts are not practical. This method of operation wastes appreciable amounts of water, but you can minimize the waste by monitoring water usage carefully and making changes in the relay when needed. As attack lines are taken out of service, you can shut down one or more of the supply lines on a multiple-line relay at each pumper. This reduces the amount of water that you must dump to maintain the desired residual pressure at the attack pumper.

The automatic relay is most useful in the early stages of a fire when conditions are changing rapidly and attack lines are being opened and closed frequently as the fire attack progresses. When usage stabilizes in the latter stages of the fire attack, it is good to modify the SOP to fit the changing conditions. At that point in the fire attack, you should have plenty of time to make needed changes, including reducing pressure settings. One way to simplify changes in the relay is to make sure that each section is hydraulically the same, so that when a pressure change is needed at one pumper (in a well-designed relay), the same adjustment can be made at the others.

An inexperienced pump operator can function satisfactorily in a constant flow relay, but a poorly trained operator can destroy the entire operation. Pump operators who might be called on to work together in a relay must know and understand the SOP and know what pressure they are to maintain. This is possible only if they participate in frequent mutual aid drills, setting up complex relays and flowing water. Constant flow relay principles will work only if a standard procedure is adopted and implemented on a regional rather than individual fire department basis.

OPEN END RELAYS

Conventional relay operations use solid connections with fire hose from the discharge of each pumper to the intake connection of the next pumper in line. The same gallons per minute flow through each section of the relay, and there must be a continuous flow to supply water to attack lines.

An open end relay uses a portable tank at the attack pumper. The relay discharges into the portable tank, instead of the pump intake, and the attack pumper drafts from the portable tank to supply water to the attack lines. The tank effectively isolates the attack pumper from pressure changes in the relay. Figure 8-5 illustrates an open end relay.

In an open end relay, the flow through the relay pumpers and supply line remains constant. The last section of supply line discharges freely into an open tank, with no restriction to the flow to cause pressure changes. If the relay is delivering water to the portable tank faster than the attack pumper is using it, the tank will overflow. Pressures in the relay then can be reduced to decrease the flow.

The water stored in the tank offers a margin of safety. For example, if a 2,500-gallon portable tank is used, the attack pumper could continue to supply two attack lines for ten minutes even if the flow from the relay were interrupted. This is enough time to get the attack crews out of the building or to a safe location before the attack lines lose their pressure.

An open relay's biggest disadvantage is that a portable tank has to be set up for the attack pumper to draft. This might require more space than you have. There is also the likelihood that the tank will overflow,

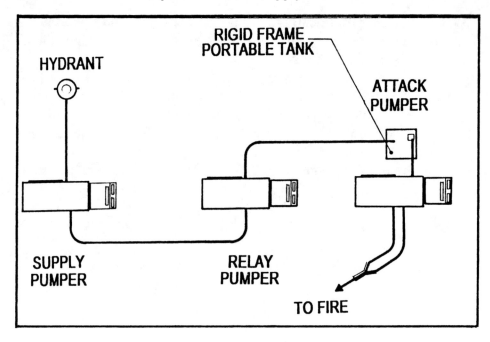

Figure 8-5. Open end relay layout.

causing a problem, especially in freezing weather. Further, the supply line that discharges into the tank could loosen and cause damage if it is not securely fastened.

In an open relay, a flexible tank can be used more effectively than a rigid frame tank because it has a built-in fitting for the supply line to make a direct connection, as well as an outlet for the hard suction hose from the pumper to use in making a connection to set up a draft.

SUMMARY

Effective relay operations are possible for most fire departments if they:

- Develop and use standard operating procedures;
- Design the relay capabilities to match the needs of a particular fire situation (considering both involvement and potential);
- Prepare, including conducting mutual aid drills to firmly establish the SOPs that will be used; and
- **Most important of all**, set up a water supply organization on every fire. Design and build your water supply—do not just let it happen.

9. Water Shuttle Operations

BEFORE YOU SET UP a water shuttle operation, ask whether fire department tankers are the most effective way to supply needed water, or whether any type of sustained water supply operation actually is necessary. Fire departments that operate regularly in areas not protected by fire hydrants should bring at least 2,000 gallons of water on the first-in apparatus. This quantity is enough to handle most single-family residences or small commercial building fires. Always keep in mind that setting up a water shuttle operation where one is not required wastes effort, and commits equipment and personnel that might be needed elsewhere.

If your water source is less than 2,000 feet away, you need a fire flow of 500 GPM or more, and if you have sufficient pumping apparatus and fire hose, a relay or large diameter hose (LDH) would better provide the needed water supply. To provide enough water to make an initial attack, tankers generally are the quickest way to transport water more than 2,000 feet and often will prove to be the most reliable.

Any fire department responsible for fire protection in areas more than 2,000 feet from an established water source must be able to establish a water shuttle quickly and maintain it over long periods of time.

Water Shuttle Advantages

Water shuttles can be used to supply water adequate to protect exposures when nearby buildings are endangered. Good water shuttle operations enable fire departments to use heavy fire streams, even master stream devices, well beyond the reach of fire hydrants. You can put a water shuttle into service quickly, when the first tanker arrives at an incident scene, often bringing under control a fire that might not wait for a relay or other complex water supply evolution to be set up.

Water shuttles require minimal personnel to operate, an important consideration given shortages of on-duty personnel, both in volunteer and paid departments. Water shuttles offer the flexibility of being supplied from hydrant systems or static sources, whichever is available and most accessible to the fire department.

Water Shuttle Limitations

Water shuttle operations are limited primarily by:

- the distance and road conditions from the water source to the scene of the incident;
- the maximum flow capability of the water source and fill pumper;
- the amount, type, size, and efficiency of the mobile water supply units that are available to haul water; and
- the time required for mobile water supply units to arrive at the scene.

However, if you organize your water shuttle to best use your available apparatus and equipment, these limitations can be overcome.

Organization

Upon his or her arrival at the scene, part of the incident commander's initial size-up should be a plan to provide the needed water. If the source for this water supply is a water shuttle, it should be an integral part of the initial plan of attack. Too often, fireground operations are set up without considering the space needed to operate a water shuttle, so that by the time tankers arrive and are ready to haul water, apparatus and hoselines have been committed to the attack, and access to and from the fire scene may already be blocked.

Some fire departments can supply more than 2,000 GPM by water shuttle, but this capability is the result of good planning and organization. Planning should start before the receipt of the alarm by identifying all target hazards and potential sources of water. After considering capacity, distance, and traffic patterns, negotiate mutual aid agreements with nearby departments and arrange automatic dispatch of water supply units. Make plans in advance with police and road maintenance personnel to obtain road condition information, and to ensure their cooperation at the scene.

Safety Concerns

Water shuttle operations can be dangerous. Personnel can be injured, and apparatus and equipment damaged. Water shuttles require a lot of movement of personnel and heavy equipment, often including maneuvers in restricted spaces and travel across highways at high rates of speed. If this is allowed to take place haphazardly, the probability of injury to personnel and damage to equipment is high. Maintaining a satisfactory margin of safety at all times should be a priority in setting up a water supply organization.

Dependability

Most firefighters in rural and suburban areas have experienced a major fire attack with multiple hoselines being supplied from different tankers. In some cases, just as the fire was nearly under control, all tankers ran out of water at about the same time. By the time the tankers had returned with another load of water, the building was lost. This happened because of a lack of organization, and a failure to control the use of the available water supply. A poorly organized water shuttle can become a massive traffic jam, and a tanker that is not moving is not transporting any water.

Efficiency

Fires have destroyed buildings because fire department hoselines were too small to handle the volume of fire that was present. Many rural fire department personnel believe they do not have enough water to use 2½"-lines or master streams for fire attack. At many major fires personnel simply use 1½"-lines to protect exposures while the fire building burns down, despite the fact that, in some instances, the apparatus and equipment are readily available to supply water sufficient to attack the fire aggressively. However, the equipment and apparatus must be used properly.

RESPONSIBILITIES

Water supply requires a separate organization at an incident scene. The incident commander's myriad responsibilities—search and rescue, exposure protection, personnel and equipment initial attack assignments, and logistical concerns during prolonged operations—demand his or her attention. There simply is not enough time to give water supply the attention it deserves, especially if a complex water shuttle operation is required. Therefore, the only solution is to appoint a Water Supply Officer (WSO) and initiate an operating structure as soon as possible after arrival at the incident scene. This operating structure should include a dump site officer, fill site officer, and traffic control officer. In Figure 9-1, an example of a water supply organization for an operating water shuttle is shown.

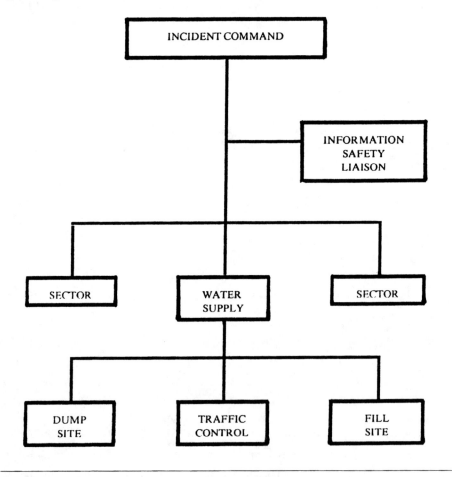

Figure 9-1. Organization Chart for water shuttle operation.

The WSO should be given full responsibility for water supply, along with the authority to use whatever apparatus, personnel, and equipment are needed to design and organize it. The WSO need not be a chief officer, or an officer at all. The only qualifications he or she needs are an in-depth understanding of water supply principles and knowledge of available resources. An officer from a mutual aid department can be assigned as the WSO if he or she knows the area and the available options. What is most important is to appoint the water supply officer immediately, before apparatus is committed and fireground operations are set up to preclude efficient water shuttle operations.

Line Officer Responsibilities

The incident commander, along with subordinate line officers, is responsible for seeing that water is used wisely. A properly organized water shuttle can haul all needed water, but not enough to waste. The "surround and drown" approach does not work when water is being hauled to the scene by tankers. The incident commander estimates how many and what types of attack lines will be needed to control and extinguish the fire, and gives this information to the WSO. The WSO then designs the water shuttle to provide the needed fire flow. Any changes in requirements during the progress of the fire attack should be communicated promptly to the water supply officer.

Figure 9-2. Officers should be clearly identified by function.

NOTE: ONCE APPOINTED, THE WATER SUPPLY OFFICER ASSUMES COMPLETE CONTROL OF WATER USAGE. NO ADDITIONAL HOSELINES SHOULD BE AUTHORIZED BY ANYONE ELSE.

The water supply officer assumes the responsibility of providing an adequate water supply to the incident scene. He or she must have the authority to decide how and when to do it.

Water Supply Officer Responsibilities

The key to a dependable water shuttle operation is controlling water usage. Since the actual flow rate is controlled by the firefighter at the nozzle, the water supply officer should base all flow calculations on the maximum flow that could be demanded by the attack lines in service. For purposes of water shuttle design, maximum flow rates are listed in Table 9.1.

Table 9.1. Maximum Flow from Attack Lines

1½″-handlines	125 GPM
1¾″-handlines	150 GPM
2″-handlines	200 GPM
2½″-handlines	250 GPM
Master stream—350 to 1,000 GPM as specified	

When short, preconnected lines are being used, it may be possible to exceed these GPM ratings. However, this practice would be the exception rather than the rule. In general, these figures provide a good basis for estimating the maximum flow rates for which the water shuttle needs to be designed.

To maintain a dependable water supply, the incident commander must be kept apprised of changing conditions; he or she, in turn, tells the water supply officer what the expected fire flow needs will be. The water supply officer should attempt to anticipate increases in consumption and be prepared before additional water supply is needed.

Once the fire flow requirements have been established, the water supply officer designs and initiates a water shuttle operation that will deliver the required amount of water to the incident scene. In order to do this, the water supply officer must:

1. Assume control of all water usage.
2. Match the supply lines and tanker transport capability to the requirements of the attack lines.

3. Assign the water supply apparatus and equipment that will be required to provide the needed flow.
4. Summon mutual aid apparatus, as required, to meet the needs of the shuttle. To provide a margin of safety, you should request one more mobile water supply unit than calculations indicate will be needed.
5. Select the most accessible fill site for tankers that will supply the needed flow rate. Assign a fill site officer to manage the fill site operation and a dedicated crew to fill tankers.
6. Select a dump site location accessible for tankers and with enough space to maneuver them. Assign a dump site officer to manage the site and a dedicated crew to take care of unloading tankers.
7. Establish the route tankers will follow in shuttling water from the fill site to the dump site.
8. Exercise overall operational control of the shuttle once it has been established.

NOTE: The shuttle design is dependent on the WSO's knowledge of each tanker's Continuous Flow Capability (CFC). The CFC rating in GPM that has been assigned after testing, according to the methodology used in Chapter 5: Testing Water Supply Apparatus, is the best indication of what to expect.

DESIGNING THE WATER SHUTTLE

Your water supply organization should be initiated as soon as you have any reason to believe the fire attack will progress beyond a single engine company's capabilities.

Call for additional tankers as soon as any indication of need arises. Design the water shuttle operation around each tanker's Expected Time of Arrival (ETA). Estimate the Needed Fire Flow (NFF) immediately. Dispatch enough tankers to supply the expected flow requirements, with at least one in reserve for a margin of safety, e.g., in case of mechanical difficulty or a delayed response. An extra tanker also can provide the needed reserve to cope with an unforeseen increase in fire volume during its early stages. Too often, fire departments find themselves playing "catch-up" with a fire that continues to expand just beyond the capabilities of their available water supply.

Figure 9-3. Pumper tanker with 1,500-GPM pump and 2,500-gallon water tank provides sufficient water for initial attack.

The water supply officer must be aware of the ETA at the fire scene of each mobile water supply unit, and adjust water usage accordingly. With adequate planning, and the prudent use of available water, each mutual aid tanker should arrive on the scene before the previous unit's supply has been exhausted.

Direct supply

The method of initial attack, and placement of the first-alarm units set the stage for the water shuttle. If the initial attack pumper is stationed in a remote location (e.g., farmyard, end of a long driveway), lay a supply line from the main road, or some readily accessible area such as a parking lot (see Figure 9-4). Arrange the dump site so that tankers can unload their water without blocking the access to the fire of later-arriving units. Consider also the necessary freedom of movement for a water shuttle at the site. If the first tanker to arrive accompanies the attack pumper to the scene of the incident to transfer its water, it may be trapped by later-arriving vehicles. If trapped, the first tanker is unable to haul water, and its Continuous Flow Capability (CFC) is lost for the duration of the incident.

If the attack pumper does not lay a line on the way in, consider laying a supply line by hand before you unload the first tanker. For water shuttles of long duration pick a readily accessible location to provide this supply line for unloading tankers. The downside of this is that

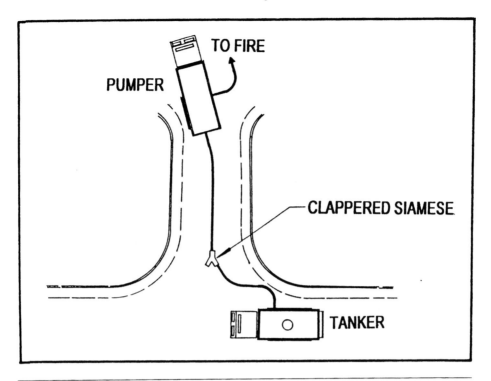

Figure 9-4. Method of direct supply from tanker to pumper.

Figure 9-5. Supply line from attack pumper allows tanker to unload at readily accessible location.

Figure 9-6. Two unloading stations allows constant supply to attack pumper.

some of the initial engine company's personnel might have to be diverted for this purpose at a critical time in the fire attack.

Most fire departments have the first tanker unload directly into the attack pumper immediately after arrival. If the tanker is equipped with a pump, no additional apparatus is needed for direct supply to the pumper. If the tanker has no pump, an additional pumper is needed for water supply; it would have to draft from the tanker to provide water through the supply line to the attack unit.

Other arrangements will work, but the most effective way to supply water directly to the attack pumper is to allow the tankers to pump water through a direct fill inlet into the tank on the pumper. If the tanker can supply water faster than the fire attack consumes it, the excess water supply will accumulate in the tank on the attack pumper, providing a reserve to continue the fire attack after the tanker is empty.

Many rural and suburban pumpers have direct fill inlets to the tank at the rear of the apparatus, but a better arrangement is to have the direct fill inlet to the tank located at, and controlled from, the normal pump operator position on the apparatus. This added convenience allows the pump operator to keep the tank full by opening and closing the tank fill valve when more water is being received than is being used to supply the fire attack lines. This prevents an overflow and wasted water.

If the attack pumper has no direct fill connection into its tank, the supply line from the tankers, or water supply pumper, can be connected to a gated intake to the pump on the attack unit. With this arrangement, the pump operator has to open the valve on the gated intake when the line is charged, and close it when the tanker is empty. Open the tank fill valve to replenish the tank after the supply line is connected to the pump intake. The amount of water that can be used to refill the tank is limited by the ability of the pump on the tanker to supply more water than the attack lines are using. The pump operator in this configuration must observe the incoming pressure closely, and open the tank fill valve only enough to divert the surplus water that is being received, not enough to reduce the flow to the attack lines.

Nurse Tanker

If the first-arriving tanker has a pump large enough to supply water directly to the attack pumper, it also can be used as a nurse tanker at the dump site (Figure 9-7). Each additional tanker can discharge its load of water into the nurse tanker as it arrives at the dump site, and return to the fill site to pick up another load of water. The nurse tanker

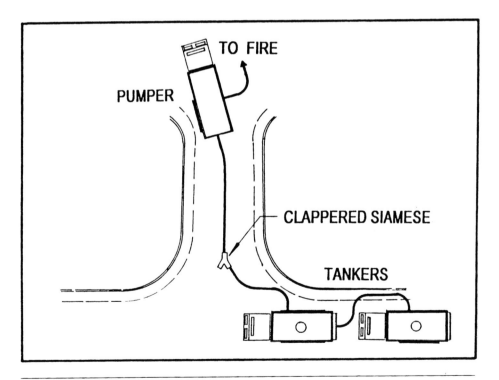

Figure 9-7. Nurse tanker provides a more reliable supply.

Figure 9-8. Nurse tanker used as reservoir for tankers to unload as they arrive.

is used as on-scene storage; it provides a constant supply of water to the attack pumper as shuttle tankers come and go.

Although using the first tanker as a nurse ensures a smooth transition from direct supply to a water shuttle, it commits a tanker for storage and confines it to the dump site, leaving one less tanker available to shuttle water. This significantly reduces the amount of water the shuttle can supply in the early stages of the fire when an adequate fire flow is most needed.

Portable Tank Operation

You can use a portable tank, instead of a nurse tanker, to store water at the dump site, and all tankers can be used to shuttle water. If a portable tank is used with a remote dump site, you will need a pumper to draft from the tank and supply water to the attack units. If the initial supply line has a clappered siamese at the supply end, the water supply pumper that is drafting from the tank can connect to the unused inlet while a tanker is still pumping into the other side of the siamese. The changeover from direct supply to the water shuttle can be made smoothly when the tanker is empty by opening the discharge valve on the water supply pumper.

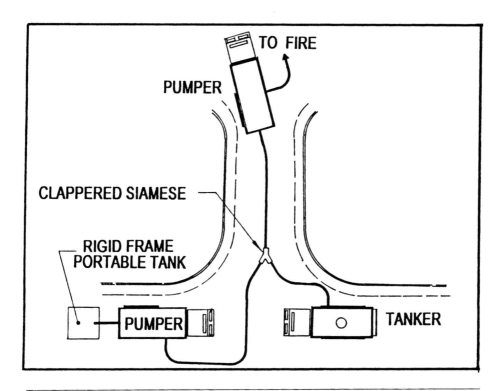

Figure 9-9. A portable tank can be used for storage instead of a nurse tanker.

To initiate a water shuttle using a portable tank at the dump site, the second tanker to arrive sets up the tank, but it's the first tanker that provides water, via direct supply, for the initial attack. After the tank is set, the tanker dumps its load of water and goes to the fill site for another load, while the water supply pumper is used to set up a draft; it then supplies water to the attack units as shown in Figure 9-9. With sufficient personnel available, the water supply pumper's engine company crew could be used as a dump site crew; they would be ready to handle the water as subsequent tankers arrive. The water supply officer could either designate the engine company officer as dump site officer, or send a more experienced officer to command the operation.

When water is arriving from the shuttle, each pump operator should divert enough water to replenish the supply in the tank on the apparatus. Fill the attack pumper's tank first, then the water supply pumper and any nurse tankers on scene, only if supply exceeds demand. When all tanks are full, they can support the attack for a short period of time if the supply is interrupted by insufficient shuttle capacity, operational difficulties, or mechanical breakdowns.

MATCHING THE SHUTTLE TO THE FLOW REQUIREMENTS

Once the IC has determined the fire flow requirements, the WSO must design a shuttle to meet the needs. In Chapter 5: Testing Water Supply Apparatus, tankers were assigned a constant flow rating based on the distance water has to be transported. You can use the same procedure to decide how many tankers will be needed for a specific shuttle, using the estimated travel time to the water supply point. To estimate the CFC in GPM of each tanker in the shuttle, it is necessary only to know how much usable water it carries and how long it will take to make each roundtrip between the fill site and the dump site. One way to determine the CFC of each tanker (this approach was used in Testing Mobile Water Supply Apparatus in Chapter 5), and the method the ISO uses to evaluate Fire Department Water Supply, is

$$\frac{V}{A + B + T} = CFC$$

where
- V = capacity of the tanker in usable water
- A = unloading time as determined by testing
- B = loading time as determined by testing
- T = estimated roundtrip travel time
- CFC = Continuous Flow Capability

Determine the usable water and expected dump and fill times in advance. Your preplan should include an evaluation of all the water supply apparatus that could be used for a specific risk, or fire management area. The WSO should be thoroughly familiar with the results of these evaluations, and a record of the CFC for each tanker that might respond should be carried on the apparatus or in the chief's car for use at a command post. If no hard data exist, substitute an educated guess, but keep in mind that the more accurate the information, the closer your estimate will be to the shuttle's actual flow. You also can make these calculations by combining the dump and fill times as handling time. Handling time is defined as the time required to handle the water at both ends of the shuttle. The revised formula would be

$$\frac{C \text{ (Capacity-Usable Water)}}{HT \text{ (Handling Time)} + TT \text{ (Travel Time)}} = GPM$$

For example, a tanker with a 1,000-gallon capacity of usable water can be loaded in two minutes and dumped in two minutes. Used in a shuttle where the roundtrip travel time is four minutes, the GPM rating could be calculated as

$$\frac{1,000}{4+4} = 125 \text{ GPM}$$

If this tanker were filled from a source that could supply only 250 GPM, the fill time would be extended to four minutes, and the handling time would become six minutes. Calculate this rating as

$$\frac{1,000}{6+4} = 100 \text{ GPM}$$

With a 2,000-gallon usable water tanker capacity, a two-minute dump time, and a two-minute fill time in the same shuttle, calculate the GPM rating as

$$\frac{2,000}{4+4} = 250 \text{ GPM}$$

Travel Time Calculations

Travel time is the time spent traveling between the fill site and dump site in a water shuttle, and it is affected by the distance to be traversed, road conditions, weather, traffic, and any other variables that tend to limit the maximum speed tankers can maintain when shuttling water. To estimate travel time, the ISO uses the formula

$$TT = .65 + (1.7 \times D)$$

where
- TT = time required for a tanker between the fill and dump sites, both directions of travel
- D = distance in miles, roundtrip
- 1.7 = time required to travel 1 mile at an average speed of 35 Miles Per Hour (MPH)
- .65 = constant for acceleration/deceleration

This formula assumes that tankers will be able to maintain an average speed of 35 MPH. This may be too high or too low. If the route is primarily limited access highways, a speed approaching 60 MPH is possible, but, when tankers must travel over narrow, poorly maintained gravel roads, 20 MPH might be too fast, potentially endangering personnel and equipment. If the expected average speed is something other than 35 MPH, the following factors should be used:

Table 9.2. Travel Time Factors by Average Speed

60 MPH	TT	=	.65	+	(1.0 × Distance)
55 MPH	TT	=	.65	+	(1.1 × Distance)
50 MPH	TT	=	.65	+	(1.2 × Distance)
45 MPH	TT	=	.65	+	(1.3 × Distance)
40 MPH	TT	=	.65	+	(1.5 × Distance)
35 MPH	TT	=	.65	+	(1.7 × Distance)
30 MPH	TT	=	.65	+	(2.0 × Distance)
25 MPH	TT	=	.65	+	(2.4 × Distance)
20 MPH	TT	=	.65	+	(3.0 × Distance)

The constant value of .65 in this formula allows approximately 40 seconds for acceleration and deceleration as the tanker approaches and leaves the fill and dump sites. Underpowered or overloaded apparatus may be slower, but experience has shown that an average speed of 35 MPH is typical for the types of road conditions typically encountered in areas where a water shuttle is generally the best means of transporting water.

Travel Time vs Handling Time

Using this formula, and assuming an average speed of 35 MPH, you can prepare a chart (similar to those included as Tables 9.3 through 9.7) that will provide a basis for water shuttle design. By relating changes in travel, fill and dump times to expected GPM capabilities, the WSO can use the chart to select the best fill and dump sites and the best route for tankers to travel to provide the most dependable water supply. The Handling Time column shown in the chart includes both fill and dump times, as determined by operational tests. Travel Time includes round-trip travel.

Table 9.3. Continuous Flow Capabilities 1,000-Gallon Tanker

Handling Time Fill Time + Dump Time	Distance from water source to scene of incident in miles (One way)						
	1	2	3	4	5	6	10
2	165	106	78	62	51	43	27
3	142	96	72	58	48	42	27
4	124	87	67	55	46	40	26
5	110	80	63	52	44	38	25
6	100	74	59	49	42	37	25
7	90	69	56	47	41	36	24
8	83	65	53	45	39	34	23
9	77	61	50	43	38	33	23
10	71	57	48	41	36	32	22
11	66	54	46	40	35	31	22
12	62	51	44	38	34	30	21
13	59	49	42	37	33	29	21
14	55	47	40	35	32	29	21
15	52	45	39	34	31	28	20
16	50	43	37	33	30	27	20

Table 9.4. Continuous Flow Capabilities 1,500-Gallon Tanker

Handling Time Fill Time + Dump Time	Distance from water source to scene of incident in miles (One way)						
	1	2	3	4	5	6	10
2	248	159	117	92	76	65	41
3	213	144	108	87	73	62	40
4	186	131	101	82	69	60	39
5	166	120	95	78	66	58	38
6	149	112	89	74	63	55	37
7	136	104	84	71	61	53	36
8	124	97	80	67	58	52	35
9	115	91	76	65	56	50	34
10	107	86	72	62	54	48	34
11	100	81	69	59	52	47	33
12	93	77	66	57	51	45	32
13	88	73	63	55	49	44	31
14	83	70	60	53	47	43	31
15	79	67	58	51	46	42	30
16	75	64	56	50	45	40	30

Table 9.5. Continuous Flow Capabilities 2,000-Gallon Tanker

Handling Time Fill Time + Dump Time	Distance from water source to scene of incident in miles (One way)						
	1	2	3	4	5	6	10
2	331	212	156	123	102	87	55
3	284	191	144	116	97	83	53
4	248	175	135	110	92	80	52
5	221	161	126	104	88	77	50
6	199	149	119	99	85	74	49
7	181	138	112	94	81	71	48
8	166	129	106	90	78	69	47
9	153	122	101	86	75	67	46
10	142	115	96	82	72	64	45
11	133	108	92	79	70	62	44
12	125	103	88	76	67	61	43
13	117	98	84	73	65	59	42
14	111	93	80	71	63	57	41
15	105	89	77	68	61	55	40
16	100	85	74	66	59	54	39

Table 9.6. Continuous Flow Capabilities 3,000-Gallon Tanker

Handling Time Fill Time + Dump Time	Distance from water source to scene of incident in miles (One way)						
	1	2	3	4	5	6	10
2	496	317	233	185	153	130	82
3	426	287	217	174	145	125	80
4	373	262	202	164	139	120	78
5	331	241	189	156	132	115	76
6	299	223	178	148	127	111	74
7	271	208	168	141	122	107	72
8	249	194	159	135	117	103	70
9	230	182	151	129	113	100	69
10	214	172	144	124	108	97	67
11	199	163	137	119	105	94	66
12	187	154	131	114	101	91	64
13	176	147	126	110	98	88	63
14	166	140	121	106	95	86	62
15	157	134	116	103	92	83	60
16	150	128	112	99	89	81	59

Table 9.7. Continuous Flow Capabilities 6,000-Gallon Tanker

Handling Time Fill Time + Dump Time	Distance from water source to scene of incident in miles (One way)						
	1	2	3	4	5	6	10
2	992	635	467	369	305	260	164
3	851	574	433	348	291	249	159
4	745	524	404	329	277	240	155
5	663	482	379	312	265	230	151
6	597	446	356	296	254	222	148
7	543	415	336	282	243	214	144
8	498	388	318	270	234	207	141
9	460	365	302	258	225	200	137
10	427	344	288	247	217	193	134
11	399	325	275	238	209	187	131
12	374	308	263	229	202	182	129
13	352	293	252	220	196	176	126
14	332	280	241	212	190	171	123
15	315	267	232	205	184	166	121
16	299	256	223	198	178	162	118

One of the decisions that has to be made when establishing a water shuttle is which fill location to use. Tankers often pass up potential water sources that would require drafting to fill at a hydrant that is more remote. If more than one source for filling tankers is available, use the source that permits the shortest roundtrip time. For example, a small stream is available for filling tankers. Although only a mile from the scene of the incident, it is not accessible to a fire department pumper and a portable pump would have to be used for filling. This would limit the maximum fill rate to approximately 200 GPM. By traveling one additional mile, a hydrant could be used that would provide a fill rate of 1,000 GPM, but require four additional minutes of travel time. If a tanker carries 1,000 gallons of usable water (as shown in Table 9.3), the five minutes needed to fill at the closest location would just about balance with the one-minute fill time at the hydrant and the four extra minutes of travel time. So, there would be very little difference in the GPM supplied from either of these fill points. However, if the tanker carries 2,000 gallons of usable water, Table 9.5 indicates that the equation would change dramatically. A 2,000-gallon tanker would require ten minutes to fill with the portable pump but only two minutes from the hydrant. In this case, traveling the extra distance to the hydrant would reduce the roundtrip time by four minutes, with a corresponding increase of as much as 25% in the GPM delivered to the fire scene. From this example, it is clear that the best design for any given shuttle depends on the water fill points, the maximum flow rate, the distance, and the tankers used to transport water.

GENERAL RULES FOR TANKER SELECTION

Tables 9.8 A and 9.8 B are comparisons of expected continuous flow capabilities of different-sized tankers over various distances; only the flow rates for filling and dumping have been changed.

Table 9.8.A

CAPACITY	ONE MILE	THREE MILES	TEN MILES
1,000	124 GPM	67 GPM	26 GPM
2,000	166 GPM	106 GPM	47 GPM
6,000	214 GPM	171 GPM	102 GPM

NOTE: Overall average flow rate during filling and dumping of 500 GPM

Table 9.8.B

CAPACITY	ONE MILE	THREE MILES	TEN MILES
1,000	165 GPM	78 GPM	26 GPM
2,000	240 GPM	135 GPM	52 GPM
6,000	374 GPM	263 GPM	129 GPM

NOTE: overall average flow rate during filling, and dumping of 1,000 GPM

These tables demonstrate two basic principles of water shuttle design:

- Long hauls require large tankers. When a 500-GPM handling rate is assumed in Table 9.8 A, the 6,000-gallon tanker supplies 72% more water over a one-mile distance than a 1,000-gallon tanker. When the distance is increased to ten miles, the 6,000-gallon tanker could supply almost four times as much flow as the 1,000-gallon tanker.
- Large tankers need high-capacity water supply fill points. If you compare the figures in the tables when the average flow rate is doubled, the capacity of the 1,000-gallon tanker increases only about 30%. The 6,000-gallon tanker, on the other hand, increased its Continuous Flow Capabilities by nearly 75% with the higher

flow rate. This difference becomes less pronounced as the distance increases because a larger portion of the total time required for a roundtrip represents travel time that is not affected by the fill or dump rate.

As a general rule in water shuttle design, smaller and more maneuverable tankers that can be filled and dumped rapidly perform well over short hauls. Larger tankers are more effective over longer distances.

Waiting Time

The formula (page 250) used for determining each tanker's capacity assumes it will be able to dump its full load at the maximum rate immediately upon arrival at the dump site. It also anticipates the unit being filled immediately when it arrives at the fill site. This is not always the case. When you encounter delays, the continuous flow capability of the tankers is reduced accordingly. Another factor has to be added to the formula that was introduced on page 250:

$$\frac{C}{HT + TT + WT \text{ (Waiting Time)}} = GPM$$

If the 2,000-gallon tanker used in a previous example had to wait four minutes to load, the Continuous Flow Capabilities would be reduced as follows

$$\frac{2,000}{4 + 4 + 4 \text{ (Waiting Time)}} = 167 \text{ GPM}$$

The additional four minutes of waiting time reduced the expected flow provided by this tanker from 250 GPM to 167 GPM. If this four-minute delay occurred on every trip, it would take three tankers to supply as much water as two could provide with no delays.

Water shuttle delays most often are caused by a lack of organization, poor planning and bad operating practices. A low fill rate will cause excessive delays in filling tankers, with resulting backups. The most common problem at dump sites is inadequate storage; tankers need to unload their full loads without interrupting or reducing the flow.

Fill site delays often occur when an attempt is made to fill two tankers simultaneously. A 1,000-gallon pumper operating from draft can fill a 2,000-gallon tanker in two minutes. If the flow is divided between two 2,000-gallon tankers, with 500 GPM going into each one, the fill time for each will be lengthened from two to four minutes. A better operating practice would be to use the pumper's entire capacity to fill the first tanker in two minutes, and send this tanker on its way back to the dump site while the second tanker is being filled.

The same type of delay can be encountered at dump sites. If the fire attack is using 1,000 GPM, a 2,000-gallon tanker can supply it for two minutes. If two 2,000-gallon tankers are dumping simultaneously, it takes four minutes to offload them, and both of them become empty at the same time. There are two exceptions to the rule that only one tanker should be dumping at a time:

1. When the fire flow is greater than the maximum dump rate of a tanker, it is necessary to dump at least two tankers simultaneously to keep up.
2. If the amount of water in storage at the dump site has been depleted, the excess flow from the tankers can be used to fill the tanks and replenish the amount in reserve.

Tankers transport water only while their wheels are turning. Any time spent waiting, either at the fill site or dump site, decreases the maximum flow rate the shuttle can deliver.

Tanker Routing

The primary concern in tanker routing is emergency vehicle and civilian traffic safety. Otherwise, the most important objective is to cut to a minimum the shuttle travel time.

Road conditions are the most important factor in route planning tanker shuttles. In addition to the distance and the type of road, weather conditions, both actual and anticipated, should be a prime consideration. Make allowances for bridges that are not heavy enough, and for narrow roads, unpaved roads, or both, and any other local conditions. Consider the size and type of tankers in your decisionmaking process.

Most highway bridges are posted with maximum weight limits that are approximately 50% of the load they can support. If some of the units in your shuttle weigh 35,000 pounds, they obviously cannot be routed over a bridge with a five-ton weight limit. Ideally, tankers should be routed so that they travel in a loop, without turning around at either end. Plan your shuttle so that tankers will not meet each other on narrow, high-crowned roads; otherwise, make it clear that loaded tankers always have precedence over empty ones. Keep the loaded tankers moving! Given a choice, tankers should travel uphill empty and downhill loaded because they are much more maneuverable empty than loaded. It is best to design the shuttle to restrict difficult maneuvering to empty tankers.

When you select your shuttle route, try to find an easy approach that requires a minimum amount of maneuvering at the fill and dump sites. You can minimize traffic problems by using more than one fill or dump site to split the traffic over two or more routes. Cut down on congestion by selecting a staging area that is removed from civilian traffic and emergency vehicles, for reserve units or those waiting to load or unload.

Moving tankers over the road is the most dangerous part of a water shuttle. Tankers are very heavy, and carry an unstable load, making them hard to control in emergency situations. Inexperienced drivers compound the problem. A common practice in some fire departments is to break in new drivers on tankers before they graduate to driving pumpers. Unfortunately, many of these people have never driven anything heavier than a subcompact car or pick-up truck, and cannot cope with unexpected emergencies when driving a heavy vehicle. Given that tankers are the heaviest and most difficult vehicles to drive in many fire departments, only the most experienced drivers should be authorized to drive them.

Traffic Control

When you organize a water shuttle, assign responsibility for traffic control, including protecting fire apparatus and tankers from civilian traffic and other emergency vehicles. Station traffic control people at all critical intersections and where tankers enter and leave the fill and dump sites. If there is no other place to set up a fill or dump site operation, block off sections of road and reroute traffic.

Consider asking local police to handle traffic control. In some areas of the country, fire police are trained in traffic control procedures and can be very helpful. In extreme situations, fire department personnel may have to assist in, or even assume direct responsibility for traffic control, until regular police units arrive on the scene.

Prolonged Operation

Sometimes water shuttle operations have to be maintained for as long as several days. This requires good planning. It also poses a logistical challenge. You might, for example, need additional tankers to replace any that are taken out of service because of breakdowns or mechanical problems. You'll also need gasoline, diesel fuel, and motor oil at regular intervals.

During prolonged operations, personnel become fatigued and fatigued workers are less alert and more accident-prone. Emergency personnel also will need food and dry clothing.

As your need for water decreases in the later stages of a fire, you can reduce the water shuttle to match the lower flow requirements. Release unneeded tankers as soon as possible to make them available for other emergencies. Because difficult and complex water shuttles often involve all of the available mobile water supply apparatus from a wide area, release apparatus as soon as possible. If the water shuttle operation will be prolonged, place mobile water supply apparatus on standby in strategic locations in the area that has been depleted.

You can reduce storage on the dump site when the Needed Fire Flow (NFF) decreases, and some of the supply lines from the water supply pumper to the attack unit can be picked up when they are no longer needed to transport the reduced flow. The dump site can be dismantled in the final stages of the incident; handle overhaul by using one or two tankers to pump directly into the attack pumper.

REQUIREMENTS FOR A GOOD WATER SHUTTLE OPERATION

A good water shuttle operation requires

- ORGANIZATION—Organization is the most important factor in operating a successful water shuttle. Organization begins when the first unit arrives at the incident scene. It involves building an operating structure that includes a water supply officer, dump site team, fill site team, and safety officers. Using mutual aid units effectively and getting the most fire attack capability from apparatus and personnel depend on the command structure.
- A GOOD SOURCE—An effective water shuttle requires a minimum fill rate of 500 GPM, but a flow of 1,000 GPM is desirable. Your filling source must be sufficient to supply the needed flow. It needs either sufficient storage, or a high recovery rate that will enable it to sustain the needed fire flow as long as required. Also, the source must be accessible. Your prefire planning should

include identifying potential water supply sources, and making whatever improvements are needed for filling tankers.
- SPECIALIZED EQUIPMENT—Most fire department apparatus and equipment have been designed for use in areas where hydrants are the customary source of water. To provide good fire protection to areas not served by central water systems, specialized equipment is required. Apparatus should be designed specifically for water supply operations. Portable tanks, quick-connect couplings, floating strainers, low-level strainers, and other equipment designed specifically for water shuttle and drafting evolutions should be readily available. Mobile water supply apparatus should be designed to function efficiently in the type of water shuttle the department normally uses. Fire departments involved in mutual aid agreements must have compatible apparatus and equipment, requiring a minimum number of adapters or emergency hookups. Consider providing a few large tankers, strategically placed throughout the area, for special situations.
- TRAINING—Since water shuttles typically are team efforts by different fire departments or engine companies, standard operating procedures (SOPs) must be adopted on an areawide basis. Training ensures a smooth operation with a minimum of confusion. To be effective, water supply training must include joint training sessions with mutual aid departments on a regular basis. A dry run of a specific preplan, including actually transporting water under the SOPs, is one of the most effective ways to evaluate water supply capabilities. Only through constant practice will all departments reach the level of proficiency needed to set up a water shuttle in a timely manner. Offer training to all firefighters, but emphasize training for drivers, pump operators, and officers.
- Document all training. The number of joint training sessions recorded is the ISO's primary consideration in developing an Automatic Aid (AA) factor during an evaluation.

When all of these requirements can be met, experience has shown that a water shuttle can provide an adequate and dependable water supply for fire attack. Fire losses will be minimized and fire officers can proceed with confidence, knowing that there will be enough water to sustain the level of attack that is needed. The biggest benefit is to taxpayers and residents of the fire department's area of responsibility; they can rest assured they will receive good fire protection, whether or not their area has hydrants in a particular location.

10. Water Shuttle Dump Site Operations

CONTROLLING WATER USAGE is the single most important aspect of water shuttle operations. Your dump site design should reflect this concern.

CONTROLLING WATER USAGE

If one pumper supplies all the attack lines it is easy to control water consumption. It is equally easy for the Water Supply Officer (WSO) to know firsthand where the water goes and to control flow changes.

A second pumper to supply attack lines on another side of a fire often is advantageous. If it is supplied by a hoseline from the discharge side of the primary attack pumper, you can still control all water usage from one point. If the demand is there supply the second pumper from a second complete water shuttle operation, independent of the first one. A second operation gives you diversity and the ability to sustain larger fire flows. The downside is that it also makes it more difficult to control water supply use and coordination.

When you rely on one pumper for all water supply, unforeseen mechanical difficulties could cause the fire attack to fail completely. Avoid this by keeping a reserve pumper at the scene.

ORGANIZING THE DUMP SITE

Dump time is the sum of the unloading, connecting, maneuvering and waiting times.

1. UNLOADING TIME—How long does it take a tanker to discharge all of its usable water? Maximum flow rate from a dump outlet is determined primarily by tank construction and pipe size, but also can be limited by dump site storage. To maintain the maximum flow rate until all usable water is discharged, you need enough storage for the excess created when the dump rate exceeds the discharge rate.
2. CONNECTION TIME—Tankers equipped with jet dumps or large gravity dumps have virtually no connection time. When water is discharged under pressure from the tanker pump to provide a direct supply to the attack pumper, nurse tanker, or to a portable tank from a remote location, you must connect hoselines before you unload and break them when the tanker is empty. To minimize this time, when each tanker arrives have hose handlers waiting and ready to connect as soon as the vehicle stops. Drivers leave their cabs at the dump site only when a pump discharges water directly, or to operate a jet dump. To avoid having someone outside of the department operate the pump, the driver may have to get out of the vehicle to operate the pump, open the discharge valves and activate the jet dump or both. With a Power Take-Off (PTO) driven pump to supply hoselines, the tanker driver engages the PTO and sets the engine to the needed RPMs without leaving his or her seat. Hose handlers open the valves and charge the dump lines. When a pump is used to supply the jet dump, the driver leaves the valves at the pump panel properly positioned to energize the jet. Engaging the PTO pump and setting the hand throttle in the cab activates the jet dump, so there is no reason for the driver to get out of the vehicle. This can shave as much as 30 seconds off normal dump time.

   Quick Connect couplings are another timesaver; they eliminate hose thread compatibility problems in mutual aid operations. Petroleum-type quick connect couplings are efficient, inexpensive, and readily available. Not as convenient are Storz type, quarter-turn, sexless couplings, but they offer some of the same advantages over conventional threaded couplings.

3. MANEUVERING TIME—Although jet dumps and large gravity dump outlets save connection time, your dump outlet must be just a few feet from a portable tank to unload. At a poorly organized dump site, maneuvering time can easily exceed time saved by eliminating hoselines.
4. WAITING TIME—Drivers of loaded tankers often must wait for tankers unloading to leave before they can position themselves to dump their load. A well-planned and organized dump site minimizes waiting time.

Dump time, fill time, and travel time determine the continuous flow capabilities (CFC) of each tanker in a water shuttle. A good location, efficient method of operation, and organization minimize time at the dump site.

ACCESSIBILITY

Dump site access is a high priority. Try to avoid your natural tendency to let the dump site develop wherever the attack pumper is located, regardless of long-term consequences. The amount of movement during a water shuttle makes it desirable to select a dump site location that is remote from the fire attack. Choose a site that keeps tanker traffic from interfering with emergency vehicle incident response and operations, but that also offers a large unrestricted area for tanker maneuvering.

DIRECT SUPPLY

If you use the water shuttle for direct supply to the attack pumper, you may have to lay hoselines only to the point where tankers connect and pump their load to the attack unit. With a clappered siamese on the end of the supply line, there is no interruption in the transition from one tanker to another (Figure 9-4) because the incoming tanker can hook up to the supply line through the unused inlet of the siamese while the unloading tanker pumps into the other one.

Try to unload only one tanker at a time, but position the incoming tanker and connect it to the unused side of the siamese on the supply line before the pumping tanker is empty. When it is empty let the standby tanker charge the supply line; there will be no interruption in the flow of water to the attack pumper.

The best way to provide a direct supply from a water shuttle to an attack pumper is to connect the supply line directly into its water tank. The shuttle keeps the tank full while the fire attack is supplied by the attack pumper through the tank to pump line. This way the peak fire flow is limited only by the attack pumper's pump capacity and the size of the tank to pump line. The water shuttle's available flow is limited by the size and length of supply lines, and the pump capacity of the tankers in the shuttle. If tankers have small booster pumps, several may have to unload simultaneously to deliver enough water to the attack pumper to achieve needed fire flow. You can connect a series of siamese fittings to provide multiple dumping stations. Figure 10-1 depicts a direct supply dump site with four dumping stations for tanker unloading.

If an attack pumper has no direct fill inlet to its water tank, you can connect the supply line to the auxiliary suction intake, and control the flow by using that inlet's gate valve. If more water comes in than is being used the tank fill line on the pumper can be opened to replace the water used from the tank for initial attack.

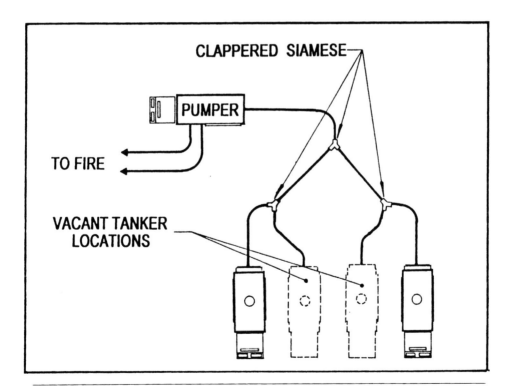

Figure 10-1. Direct supply to attack pumper from four tanker dumping stations.

Figure 10-2. Using large diameter hose to attack pumper with gated wye for tanker unloading lines.

Direct supply works best if the initial attack pumper has a high-capacity water tank (preferably 1,000 gallons or more). Some suburban and rural departments use pumper tankers with water tanks up to 2,500 gallons for heavy attack. Many such large pumper tankers have multiple fill connections directly into the water tank, designed for direct supply from a water shuttle.

MINIMUM STORAGE REQUIREMENTS

Maintain an uninterrupted supply to the fire attack by storing as much water as possible at the dump site. In fact, store at least three times your expected maximum fire flow. If the attack pumper's tank holds 1,500 gallons of usable water, and you are attacking with two 2½"-lines whose maximum expected flow is 500 GPM, the attack pumper would run dry in three minutes. Even in well-organized water shuttles periods of three minutes or more when no tankers unload are not unusual. Extra dump site storage can be a safeguard against mechanical problems, operational difficulties, traffic congestion, and other delays in loaded tankers returning to the dump site.

The direct supply method uses minimal space at the dump site, and is the simplest operation. You will still need extra space at the site if the attack pumper's water tank holds less than 1,000 gallons, or you are using multiple heavy attack lines. Use a nurse tanker or portable tank for extra storage.

NURSE TANKER

If you use one of the tankers as a nurse for the attack pumper, leave it at the dump site as a place for each shuttle tanker to unload (Figure 9-7). As each tanker unloads and returns to the fill site for another load, the nurse tanker provides a constant supply to the attack pumper. A nurse tanker with a pump can deliver the water stored in its tank directly to the attack pumper, but the pump can be used only if its capacity is high enough to supply the maximum fire flow. If a nurse tanker has no pump or one that is too small, assign a pumper to draft from a large outlet on the tank, and act as a water supply pumper. But be certain that your dump site location is large enough, and its terrain suitable, to position the water supply pumper to draft from the nurse tanker through a hard suction hose.

Terrain is a factor in how much usable water a nurse tanker can supply. Water accumulates at the low end of an unlevel apparatus tank and some will remain in the tank when the pump loses its prime, leaving it unavailable for fire attack. This happens most often when a tank has just one outlet, and the tanker is positioned so that its outlet is on the high end of the tank. To get more usable water, plan the dump site to position the nurse tanker where its tank is level or the outlet is at the tank's low end. Generally, about 90% of a tanker's total capacity is usable from a nurse tanker if the tank is nearly level.

Very large tractor trailer tankers are common in some areas. Typically without pumps, they may, at best, have a small separate engine driven or PTO booster pump to unload. Large tankers with low-capacity booster type pumps function more efficiently as nurses if you use a pumper to draft from them (Figure 10-3). Very large tankers are clumsy, hard to maneuver, and not well suited for shuttle operations. On the other hand, at large dump sites, they make ideal nurse tankers.

You'll find nurse tanker operations easier if your tankers have functioning water level gauges. The WSO checks the gauge to continuously monitor the amount of water in storage, and to adjust water usage, if necessary.

As a general rule, larger nurse tankers increase the water shuttle supply's reliability. However, if you use a large tanker you could find yourself without enough shuttle tankers to maintain needed flow.

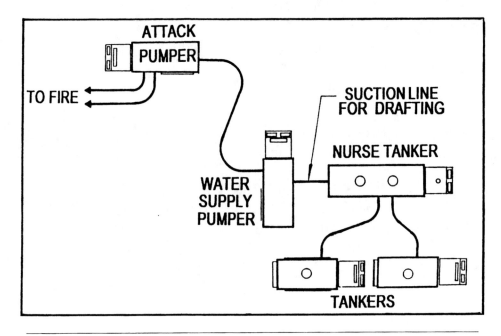

Figure 10-3. Water supply pumper drafting from nurse tanker to supply attack pumper.

Figure 10-4. Water supply pumper drafting from nurse tanker.

PORTABLE TANK OPERATIONS

A portable tank used to store water at the dump site can take the place of the nurse tanker, and release all tankers to haul water. It's also a way for tankers that lack major fire pumps to unload through large gravity dump outlets or jet dumps at very high flow rates. There are two types of portable tanks: rigid and flexible. The most common have rigid metal frames, are always the same size and shape wherever they are set up, and need space to be used effectively. Flexible tanks tend to be round, but have no fixed shape and conform to the contours of the terrain and, to a certain extent, the available space.

Rigid Frame Tanks

Rigid frame portable tanks come in different gallonages and dimensions. A typical size is 10-feet square, is 30" high, and holds approximately 2,000 gallons of water. Standard rigid frame tanks are square, although your department can custom order other shapes.

Figure 10-5. Dump site using nurse tanker.

Flexible Storage Tanks

Flexible storage tanks are lighter and require less storage space on apparatus than rigid frame tanks. A 2,000-gallon flexible tank can be folded to fit a 3' × 2' carrying case. It is possible to add one to an already heavily loaded attack pumper that has no space for a rigid frame tank.

Flexible tanks unfold flat, and have floating collars that rise as the tank fills with water (Figure 10-6). Since the tank rim floats on top of the water, you can avoid many of the problems rigid frame tanks exhibit on uneven terrain.

Flexible tanks have a five-inch suction outlet with National Standard Threads (NST) welded into the fabric at the bottom of the tank. You can equip this suction outlet with the type of fittings needed to match the pumper you use to draft. However, only a pumper with the proper adapters to connect to the tank can use it, whereas you can use a rigid frame tank effectively with any pumper that has a floating or low-level strainer to draft from it. Also, the suction outlet's location near the bottom of the tank means that a pumper cannot empty the tank completely. The six inches of water typically left in the tank when the

Figure 10-6. 1,500 gallon capacity Flexible tank.

Figure 10-7. Flexible tank ready for storage.

pumper loses its prime reduce the percentage of usable water. Flexible tanks also have a five-inch inlet with National Standard Threads (NST) fabricated into the tank. The flexible tank is designed to be filled by supply lines from the pump on the MWSA. The right combination of siamese and/or wye fittings means that you can use an almost unlimited number of pressure dump lines to fill this type of tank without special fill devices or hose holders. Fill openings often are equipped with Storz quarter-turn fittings so the fire department can use large diameter hose for filling it. You can use multiple supply lines with different types of hose if there is a short section of large diameter hose from a manifold to the tank. A good practice is to install an adapter with a gate valve at this pressure fill fitting; then you can make changes, and add other fittings, without water escaping and interrupting the water shuttle.

Flexible tanks are problematic when you use large gravity dump outlets or jet dumps, because with the shape of the tank (Figure 10-6), and the smaller opening at the top, it is more difficult to maneuver tankers into position to dump directly into the tank. Initially, filling is more difficult, and you need extra help until there is enough water in the tank to allow the flexible collar to float and give the tank shape.

Figure 10-8. Pumper with incompatible threads on suction hose drafting from flexible tank while tanker dumps from gravity outlet.

PROTECTING THE TANK

A tank's size, and the need for tankers to get close enough to dump directly into it, make its location important for a smooth-working dump site. The tank needs to be in a relatively smooth and level location because its liner is fabric and susceptible to damage from pointed rocks, gravel, or other debris. Sweep the area first, eliminating pointed rocks or other damaging objects, before you position the tank. More practical is to put a salvage cover or other protective covering between the tank and the ground.

USABLE WATER

Put a rigid frame tank on level ground to avoid reducing its capacity. Otherwise water overflows on the low side of the tank before the tank is filled to capacity. The steeper the grade, the less water the tank will hold. For example, if the low side of a rigid frame tank is 6"-lower than the high side, it holds 10% less water. In a typical 2,000-gallon tank, this difference would reduce the usable water to 1,800 gallons. In larger tanks, the difference is greater. On slightly uneven ground, you can block up the frame on the low side of the tank; the fabric conforms to the new shape, and the tank's full capacity is preserved. But if the difference in elevation is more than a few inches, the tank could fall off the blocks, and damage the frame or liner.

Sloping ground affects flexible tanks, but not as much. The tank's floating collar remains level, and the shape of the tank distorts to match the terrain. The amount of water left in a folding tank when the level drops enough to cause the pump to lose its prime is a factor in determining the amount of usable water. It is impossible to draft all of the water from a folding tank, because as the depth of the water over the strainer decreases, a whirlpool develops. Eventually, the strength of the whirlpool allows air to enter the suction hose. Then, the discharge pressure fluctuates. Finally, enough air enters the suction hose to cause the pump to lose its prime, interrupting the flow. However, with a properly positioned low-level strainer or floating strainer, the pump can continue to draft until the water depth drops as low as three inches. The amount of unusable water left in the tank varies with the pump's flow rate: the higher the flow rate, the more pronounced the whirlpool, and the sooner the pump loses its prime. Terrain also affects the amount of unusable water left when the flow is interrupted. If you use the pumper to draft from the low side of the tank, you can remove more water before the suction hose draws air than if the strainer is positioned on the tank's high side.

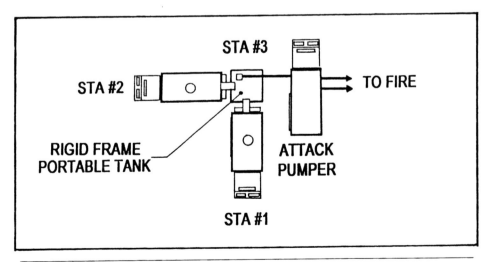

Figure 10-9. Attack pumper drafting from portable tank in parallel position.

Poor dump site design decreases the amount of usable water in the tank. When tankers dump into a folding tank at high flow rates, it generates a great deal of turbulence. This tends to happen when tankers use large gravity dump outlets or jet dumps, or when tankers try to unload faster by applying too much pressure to pressurized dump lines. Position the suction line from the water supply pumper in the tank so that it keeps the strainer as far away from the tanker-dumping area as possible (Figure 10-9). Turbulence can allow air to be drawn into the suction hose as the level drops; this causes the pump to lose its prime while there is still a substantial amount of water left in the tank.

POSITIONING THE TANK

Locate the portable tank where a water supply pumper can draft from it without blocking shuttle tanker access. In some cases, the attack pumper can draft directly from the portable tank, eliminating the need for a separate water supply pumper. When used for water supply, the location of the attack pumper limits the dump site's flexibility and your ability to expand it into a "High-Capacity Water Shuttle" as the fire progresses.

Figure 10-10. Pumper drafting from parallel tank with strainer positioned for least turbulence.

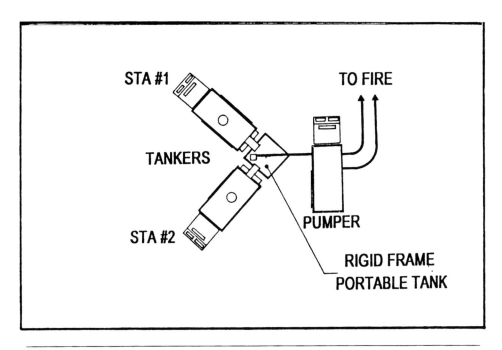

Figure 10-11. Attack pumper drafting from tank in diamond position.

Figure 10-12. Pumper drafting from tank with floating strainer floating free.

Try to arrange the tank so that shuttle tankers can dump from more than one dumping station. Figures 10-9 and 10-11 show two different ways to do this. While the normal procedure is to empty one tanker completely before the next one begins to dump, there are times when you must unload several tankers simultaneously. Whenever tankers in a shuttle have a maximum dump rate that is lower than the needed fire flow, more than one tanker must dump at the same time. With multiple unloading stations, one tanker can dump while another gets in position, minimizing delays and shortening waiting time.

SPECIALIZED EQUIPMENT

You need specialized equipment to use portable tanks effectively. For example, if a supply pumper uses a standard barrel type strainer for drafting, it can begin to develop a whirlpool, and lose its prime, when the level in the tank drops to 8" or less. Since the average folding tank is 30" deep, the remaining 8" of water represents more than 25% of the tank's capacity. With a floating strainer, or a low-level strainer specifically designed for folding tank operations, you can draw the water level down to approximately 3" with 10% versus 25% of the total capacity unusable.

Figure 10-13. Whirlpool created by high flow when level in portable tank drops below three inches.

Strainers for Folding Tank Operations

Low-level strainers or floating strainers work equally well in folding tanks; the choice is between economics or logistics. The low-level strainer is smaller, easier to handle, and takes up less space on a pumper. The floating strainer is bulkier, heavier, and takes up more storage space, but you can use it to draft from a static source, making it essential for pumpers in rural areas.

Although not designed for drafting from ponds or streams, low-level strainers can be used for this. With an adapter, you can use a low-level strainer to draft from a pond or stream, but these adapters are not readily available. Some fire departments use various materials to construct floats to elevate low-level strainers: one fire department carries a truck innertube, inflates it from the air brake system on the apparatus, and uses it as a float with the low-level strainer (Figure 10-14). If you add some type of external float, you can use the low-level strainer to draft from static sources and for portable tank applications.

Older types of floating strainers function as well as a low-level strainer in a portable tank, but newer versions need to be modified for safe use. The original floating strainer was round on the bottom like some current models. Others are flat on the bottom where the water enters the suction hose. Flat strainers work for drafting with less water than round-bottomed ones, but they pull the fabric from the bottom against them, forming a seal when the tank is nearly empty. The vacuum in the suction hose can damage the bottom of the tank or the

Figure 10-14. Innertube used to create floating strainer our of low level strainer.

strainer, because a high vacuum develops when the tank liner fabric blocks the intake holes in the strainer. To use a floating strainer with a flat intake, you must modify it to maintain a distance of at least one inch between the bottom of the strainer and the tank's fabric. Figure 10-16 shows a modified floating strainer. One manufacturer sells a riser plate you can attach to the floating strainer to maintain the necessary clearance.

Figure 10-15. Floating strainer will not lose prime until level drops below three inches.

Figure 10-16. Floating strainer modified for use in portable tank.

Figure 10-17. A shovel, garbage bag, or floating beach ball can be used to minimize whirlpool effect when using barrel strainer.

Figure 10-18. Homemade low-level strainer.

GATED INTAKES

You need a gated suction intake on the water supply pumper, the attack pumper, or both to get a dependable supply of water from a water shuttle. With a gated suction inlet the supply pumper can use the water in its tank when the water in the portable tank gets low. Close the gated intake and open the tank to pump valve, and you can supply water from the pumper's tank to sustain the attack until the next tanker arrives, especially important if your attack pumper is drafting directly from the portable tank. Initiate the fire attack with the water in the attack pumper's tank. After you set up and fill the portable tank, connect the hard suction hose to the gated intake without interrupting the water supply to the attack lines. When you change over to an external supply in the portable tank, open the valve on the suction intake slowly while you close the tank to pump valve in a coordinated movement. You can complete this transition with little or no interruption to the fire attack. Do not make the mistake of using the gated auxiliary 2½''-suction to draft from the folding tank. Although using this smaller intake with a ball valve to control it makes connecting the suction hose easier and the transition simpler, it limits the maximum flow from the water shuttle to no more than 250 GPM.

If the attack pumper has no gated front or rear suction, install an external butterfly or gate valve. There is an externally mounted valve that can make the transition from the tank on the apparatus to drafting from the portable tank automatically without interrupting flow. The automatic suction valve, installed on a large inlet to the pump, can be used three different ways: opened manually, closed manually, or left in the floating or automatic position (Figure 10-20). When you set this valve to the automatic position, you can make the changeover by simply closing the tank to pump valve slowly. As you close the tank to pump valve, friction loss in the tank to pump line increases, creating a vacuum at the pump's intake. The vacuum causes the valve to open slowly enough to exhaust the air from the suction hose and draft water from the portable tank without interrupting the flow to the attack lines.

HOSE HOLDERS

When tankers unload by discharging under pressure from a fire pump, the discharge hose has to be fastened to the portable tank securely enough to prevent equipment damage or injury to personnel.

Figure 10-19. External butterfly valve on front suction of a pumper.

Figure 10-20. Automatic suction valve and lightweight flexible suction hose on a pumper.

Figure 10-21. Homemade diffuser for pumping into a portable tank.

What is dangerous is the reaction that is created when a large volume of water flows from the unrestricted opening on a hoseline. There are commercial devices available to secure the hose to the portable tank, but some have failed under pressure, damaged apparatus, and exposed bystanders to potential injury.

Some of the most effective appliances designed for pumping into a portable tank have been fabricated by fire department members. The key to discharging large volumes of water safely is to disperse the streams where they leave the hoseline. This reduces the reaction. One of the simplest devices for this is shown in Figure 10-21. This device was constructed by welding a plate across the opening of a 2½″-coupling, and cutting openings on all four sides. The opposing streams that flow from this device cancel the reaction, and only a minimum amount of reaction is transmitted to the supply line. A pipe with a 90-degree bend and a tee extending to the bottom of the tank, and with a female coupling on the other end, will serve the same purpose.

You can use equipment originally purchased for other purposes to pump into a folding tank. Use a low-level strainer with pressure dump lines to diffuse the water in all directions, and reduce the force of the reaction as the water leaves the openings. Figure 10-22 shows a low-level strainer being used this way. Or, use a barrel strainer, in conjunction with 2½″-hard suction hose, to discharge water in all directions; disperse so many opposing streams that the overall reaction is negligi-

Water Shuttle Dump Site Operations 283

Figure 10-22. Using a low-level strainer to pump into a portable tank.

Figure 10-23. A 2/12"-barrel strainer can be used as a diffuser for pumping into a tank or filling tankers.

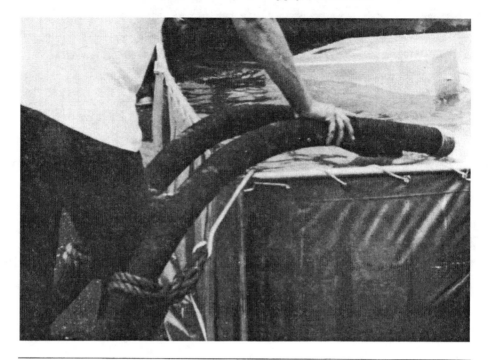

Figure 10-24. Hard suction hose used as a nozzle to pump into a portable tank.

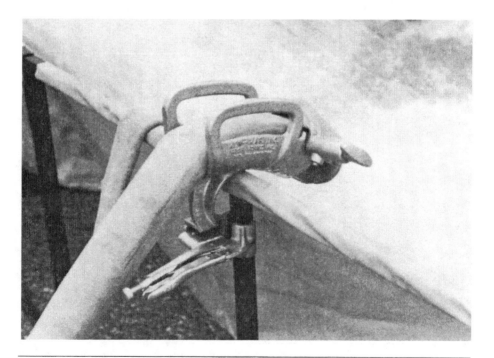

Figure 10-25. Commercial hose holder.

ble. Figure 10-23 shows the use of a barrel strainer as a diffuser. One advantage of using a section of 2½"-suction hose at the end of the pressure dump line is that you can secure it easily with a rope hose tool or other device. You also can secure pressure dump lines by using a ladder tied to the frame of the tank as a holder for a number of lines, fastened so that they discharge into the tank. One rule never to be violated is

> ALWAYS FASTEN A PRESSURE DUMP LINE TO THE TANK SECURELY BEFORE YOU PRESSURIZE IT.

DUMP EXTENSIONS

Among the devices used to extend the dump outlet on tankers (and reduce maneuvering time and simplify the operation) are discharge tubes to connect to 2½"-discharges on the tanker to take the place of pressure fill lines that supply the folding tank. These have been constructed by fire departments to use with either rear discharges or side discharges, depending on dump site traffic patterns and compatibility with other shuttle tankers. These dump tubes work best with high-capacity fire pumps; the maximum dump rate is limited by the tank-to-pump flow capability of the tanker which is dumping. Figure 10-26 shows a pumper tanker that is discharging through dump tubes connected to two rear discharges.

Jet dumps work well with dump extensions. In some areas, a standard arrangement is to use an adapter with a jet dump that allows you to use irrigation pipe to extend the dump outlet and reduce maneuvering time. This irrigation pipe comes in 5- and 10-foot sections of straight pipe or 90-degree-bend pipe you can configure for any dump site. With a jet dump, water velocity as it leaves the dump opening is high enough to keep the dump rate from being significantly affected by adding an extension. Some departments construct homemade devices of PVC pipe, heating duct, or hard suction hose for this purpose. You can use a section of hard suction hose for a dump extender if the dump outlet is the same size, and has the same threads the pumper uses for drafting. Or, vary the arrangement by cutting off a section of damaged hard suction hose to create an extension that is long enough to reach the tank without excessive maneuvering, yet small enough for one person to handle safely and quickly. If a piece of flexible suction hose is used for an extender, it can be used to dump on either side of the apparatus.

Figure 10-26. Pumper tanker pumping into a portable tank through dump tubes on rear discharges.

Figure 10-27. Section of irrigation pipe used as jet dump extension.

Water Shuttle Dump Site Operations

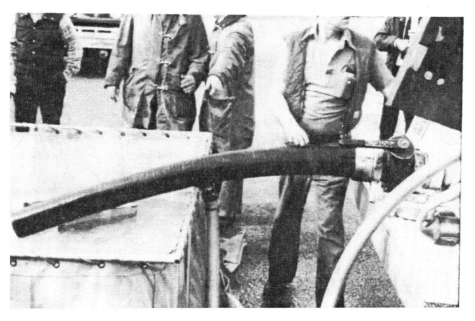

Figure 10-28. Left, adapter on jet dump outlet to use irrigation pipe extender.

Figure 10-29. Below, suction hose used to extend dump outlet.

Some large gravity dumps have built-in telescoping extensions, while others have extension tubes that are the same size as the dump, and can be fastened to the outlet, often built with 90-degree bends. Some tankers have large gravity dumps to the side of the vehicle, instead of, or in addition to, the one at the rear. On others, the dump tubes are recessed into the body on each side and can be extended by air pressure.

Whatever type of extender you have, your primary objective is to reduce the maneuvering time at the dump site. Pressure dump tubes could save the time it takes to extend fill lines to the folding tank; this simplifies the dump site operation.

USABLE WATER IN PORTABLE TANKS

The better designed and organized the dump site, the higher the percentage of usable water in the portable tank. A good rule in dump site design is to consider 75% of the water storage capacity usable. This is a conservative, but safe, estimate.

DUMP SITE ORGANIZATION

To keep time at the dump site as short as possible, shorten the roundtrip time, and deliver the maximum GPM from the shuttle, you need an organized dump site with a clearly delineated chain of command. Below are specific responsibilities for the dump site team members.

Dump Site Officer

- Assigns specific jobs to all dump site team members and clarifies their responsibilities.
- Oversees the entire operation. Tells the pump operator how much pressure is required and the rate of flow that can be expected.
- Communicates constantly with fill site officer and water supply officer.
- Informs both when difficulties are encountered or shuttle cannot supply needed fire flow.

PUMP OPERATOR

- Operates the pump on the water supply pumper that is drafting from a portable tank or nurse tanker, or the pump on the nurse tanker when it is delivering water directly to the attack units.
- Maintains proper pressure on each attack line and makes sure enough water is available to supply them when the attack pumper also serves as the water supply pumper.
- Switches back and forth from the external supply to the tank on the apparatus as conditions change.

(Since the entire fire attack depends on one pumper and its operator, that operator must be trained and experienced in water shuttle operations.)

Hose Handlers

(The number of hose handlers needed varies with the number of tankers expected to dump simultaneously and the method they use to unload.)

Figure 10-30. Directing tanker into position to dump.

- When jet dumps or large gravity dumps are used, connect and disconnect all hoselines from tankers at the dump site when pressurized dump lines are used.
- Wait, with proper hoselines ready, as each tanker arrives.
- Connect dump lines as soon as tanker stops.
- Hand tighten couplings (do not use a spanner wrench—it causes delays in making and breaking connections, and minor leaks are less important than extra time spent at the dump site).
- Eliminate delays caused by drivers or other personnel getting out of vehicles to make their own connections.
- Determine best way to unload each tanker the first time it arrives at the dump site.
- Understand differences in hose threads or couplings that could cause problems if tankers try to hook up to wrong lines.
- Disconnect hoselines as soon as tanker is empty.
- Signal drivers to leave immediately and return to fill site for more water.

Vent Crew Members

- Open hatch on top of the tanker if tank vents are not adequate. (Vents must provide enough air to prevent vacuums from developing as water is removed from the tank. There is little need for vent crews at most water shuttles, even less as outdated equipment is replaced.)
- Check tankers unfamiliar to the dump site officer the first time they unload to see if the vents are adequate. (If they are not, tank damage could occur if tanker is unloaded at the maximum flow rate.)
- Check vents by releasing the hatch covers, and opening them a crack. Air rushing through the narrow opening indicates that regular tank vents are inadequate, and a pressure differential is being created. Vent crews make sure hatches are open each time an inadequately vented tanker unloads; they also pass this information on when they are relieved. Keep vent crews off apparatus until it stops. When tankers are empty, the traffic control officer verifies that the vent crew is clear of the vehicle before signaling the driver to leave the dump site.

Traffic Control Officers

- Direct tankers approaching the dumping stations to the proper position as they arrive.

Safety Officer

- Oversees all activity and takes action to eliminate unsafe practices and safety hazards.
- Directs tankers that are backing up to the dump site.
- Installs wheel chocks to keep tankers from moving.

CAUTION: NEVER GET BETWEEN A TANKER THAT IS BACKING INTO POSITION AND A PORTABLE TANK.

Tankers must get very close to tanks to dump, and if a driver loses control, personnel caught between the tank and the tanker can be injured seriously. If it is necessary to get between the tank and the tanker to open the dump valve, make sure the vehicle has stopped and wheel chocks are in place.

If there is a personnel shortage, the Dump Site Officer may need to function as the safety officer. This is a very important function. It is better if the Safety Officer has no other responsibilities. Appoint a dedicated safety officer as soon as possible.

Since the dump site crew has to function as a team, designate a mutual aid engine company to set up and operate the dump site. Include an officer, a pump operator, and enough firefighters to handle the rest of the jobs that must be done. The biggest advantage of using a complete engine company to operate the dump site is that the members tend to belong to the same fire department and are accustomed to working together in various evolutions.

OPERATING A DUMP SITE

When tankers arrive at the dump site for the first time, the dump site officer has to decide how best to unload them. Tankers that depend on jet dumps or large gravity dumps for unloading must get close enough so that the portable tank can catch water as it comes out of the dump opening. Tankers with pumps can be unloaded at a distance from the portable tank or nurse tanker. Lay hoselines from the storage tank to a remote location so they can unload without interfering with the dumping tankers. Fasten these lines securely in place, and leave them in position for the duration of the shuttle. Tankers with small pumps and small gravity dump outlets can unload most quickly by dumping and pumping simultaneously. Tankers with large gravity dumps and a major fire pump are more flexible. They can use the large dump if the portable tank is accessible, or use the pump to unload at a remote location if the dump site gets crowded. Establish traffic patterns that allow each tanker in the shuttle to unload in the most efficient manner.

It is best to allow only one tanker at a time to dump. This way each unloads in the shortest possible time, and can go for more water while the next tanker in line unloads. If your dump site officer does not control the operation, you could have three or more tankers dumping simultaneously, each gated down to reduce the dump rate to keep the storage tank from overflowing. If this happens, all could be empty at about the same time, and you would have to withdraw attack lines and suspend the attack while your tankers go for more water.

When tankers dump, do not let them dump completely because, as the water level in their tanks drops, the head pressure also drops, decreasing the rate of flow from the gravity dump. It takes about the same amount of time to discharge the last 20% of water in a tank through a gravity dump as it does to discharge the first 80%. So, there is a point of diminishing return where the time it would take to drain the tank completely is too long to justify the extra water it would supply. You know this point is approaching when the stream breaks or no longer comes out of the dump with enough velocity to carry all of the water to the tank in a solid stream. A gravity dump opening that no longer flows a full stream of water is another indication. If you are using a jet dump, the shutoff point comes when only the water coming out of the jet is discharging through the dump opening and the pipe is no longer full. With a tanker using a pump to unload, consider the cutoff point to be when the pump first starts to cavitate. While you could coax a few more gallons from the tank, the wasted time is not worth it.

Use the higher-GPM capability tankers as efficiently as possible. Consider establishing separate dumping stations for high- and low-capacity tankers to keep the low-capacity tankers out of the way of the high-capacity tankers. There is no requirement for each unit to take its turn; rather, the objective is to use all available equipment to get the most fire attack water. Use the most efficient tankers to their full potential and give them priority when they are waiting to dump.

11. Water Shuttle Fill Site Operations

WHILE IT MIGHT SEEM that operating a dump site is more complicated than operating a fill site, the reverse is often true. Experience has shown that it is the fill site that is most often responsible for limiting a water shuttle's maximum flow, often because the water source is remote from the incident scene. This means that the fill site operation receives little attention from the incident commander and may have to function with limited personnel.

FLOW LIMITATIONS

Don't overlook the fact that a water shuttle cannot deliver more water to the incident scene than the source can supply. The maximum flow you can get from a shuttle using a 1,000-GPM pumper to draft and fill tankers is 1,000 GPM. This is a theoretical number, and assumes that:

- There will always be a tanker in position to load;
- Each tanker will accept water at a rate of 1,000 GPM;
- The source will provide a constant flow of 1,000 GPM; and
- There are no delays in beginning to fill tankers.

This never happens. Based on experience, it is difficult to maintain a continuous flow of more than 75% of the peak flow rate from the water supply pumper at the fill site. When the fill site is using a 1,000-GPM pumper to draft, the practical limit of continuous flow you can expect from the water shuttle is 750 GPM. Even at an average rate of flow, 75% of the peak flow is difficult to sustain without a good source, a well-designed fill site, and a smooth-running organization.

SELECTING A SOURCE

When you design a water shuttle, should you use a hydrant to fill tankers or a pumper at draft? There is no universal answer to this question. You have to ask, "Which fill site will allow tankers to make a roundtrip and return to the incident scene in the shortest possible time?"

MAXIMUM FLOW CAPABILITIES

Your first fill site consideration is maximum flow rate, which is limited by the water source capacity, rated capacity of the pump on the fill pumper, size, number, and length of the fill lines, and the construction of the tankers being loaded.

HYDRANTS AS A SOURCE FOR FILLING TANKERS

Generally, the water distribution system limits the maximum flow rate from a fire hydrant, and the available flow from a water system varies widely from hydrant to hydrant. Even hydrants physically located within a block of one another can have drastically different flow rates. If you use a pumper to boost the residual pressure at the hydrant, a higher fill rate is possible, but the peak flow rate still is limited by the amount of water the system can deliver to the hydrant being used.

Some rural systems have small distribution mains that cover wide areas. The flow rate from these systems is strictly limited. One rural system in South Carolina has a maximum flow rate of less than 300 GPM from most of its hydrants. You can boost your peak flow when you fill tankers from a low-flow hydrant by setting up a portable tank. Then, instead of connecting from the hydrant to the tanker, water is discharged from the hydrant to supplement the water in storage.

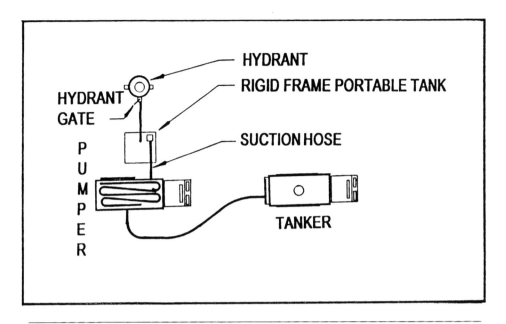

Figure 11-1. Using a low flow hydrant and portable tank to attain a maximum fill rate for tankers.

Figure 11-2. Portable tank supplied from low flow hydrant.

The capacity of the water supply pumper at the fill site also limits the rate of flow, but a high residual pressure from the hydrant enables the pumper to deliver a rate of flow well in excess of its rated capacity. As a general rule, a pumper will be able to deliver as much as 10% above its rated capacity when operating from a hydrant with 20 PSI residual pressure.

In addition to the flow rate, the amount of water stored in the system is important. When a fire flow in excess of 1,000 GPM is required, it is possible to completely drain a small water system before the fire is extinguished. The water supply officer (WSO) should notify the operator of the water system when large amounts of water will be withdrawn for filling tankers. All of the pumps should be set for manual operation, and the maximum production capabilities of the system can be used to supplement the water in storage.

DIRECT FILL OPERATION

What is the quickest way to put a fill site into operation? Connect the fill lines directly from a hydrant to the tanker. Install external valves on the hydrant outlets and use them to control the flow and to save time opening and closing the hydrant each time a tanker is filled.

Figure 11-3. Soft suction hose used for direct fill from hydrant.

Some tankers have 4½"-tank fill connections with National Standard Threads (NST) on the coupling. Others have 5"-tank fill inlets with Storz fittings that allow them to be filled with large diameter hose. If you connect a 15-foot section of soft suction hose or LDH directly from a hydrant to this inlet, you can get fill rates as high as 1,500 GPM from a typical water system. Other tankers have gravity or jet dump outlets with NST connections that can be connected to the hydrant with soft suction hose. Because dump outlets are constructed to allow water to flow out of the tank at very high rates, use them to fill the tank at the same rate by reversing the direction of flow. At high flow rates overly restrictive baffles can keep water from moving freely from compartment to compartment inside the tank during filling. If individual compartments fill prematurely, water may be forced out the overflow, depending on its position, before the tank is actually full. The advantage of using a dump outlet at the rear of the tank to fill when the vent/overflow is at the front is that water will not flow from the vent until the front compartment is full. In fact, it will not happen until the entire tank is full.

If hydrant-to-street distance precludes reaching the tanker with a 15-foot section of soft suction hose, extend your reach with a section of hard suction hose or LDH. Couple the soft suction hose to

Figure 11-4. Using a section of hard suction hose for an extension from a hydrant to a soft suction hose for filling tankers.

the end of the hard suction; you can connect easily to the tanker. Leave both sections coupled and in place while you wait for the next tanker to arrive.

With the 2½"-direct fill inlets on most tankers, and depending on hose length and residual pressure at the hydrant, it is possible to achieve a flow rate as high as 1,000 GPM filling through a single 3"-supply line. Tankers without direct fill connections may have to be filled through the suction inlet of the fire pump, either with the tank fill valve or the tank to pump line. The tank fill line is usually 1½"-pipe or smaller, too small to fill the tank fast enough to be effective in a water shuttle. The tank to pump line generally is much larger; you can expect a fill rate of at least 500 GPM when you open the valve and water flows backwards from the pump into the tank. Most pumps have a check valve in the tank to pump line to keep water from entering the tank under pressure and damaging it. If your tanker has this construction, do not try to use the tank to pump line for filling. When tankers have dual tank to pump lines, one is probably equipped with a check valve, while the other is not. Mark the valves clearly so the operator will know which tank to pump valve should be used for filling the tank from the pump intake connections. When you fill through 2½"-inlets, either 2½"- or 3"-lines can be connected from the 2½"-outlets on the hydrant. As a time-saving measure, connect external quarter-turn or gate valves to the hydrant outlets to control the flow. Leave the hydrant open all the time. Since hydrants have to be fully opened each time they are used, and then fully closed when shut down, time is lost opening and closing them for each tanker. You also can supply 2½"- and 3"-supply lines by using a gated wye with the 4½"-female coupling on the large outlet of the hydrant, and connecting the supply lines to the gated 2½"-male couplings.

If valves to fit the 4½"-pumper outlet on hydrants are not readily available, the hydrant operating valve must be used to control flow. This is why it is not practical to use the same hydrant to supply large and small fill lines. It is simpler to use two different hydrants for filling tankers with different-sized inlets operating in a shuttle. Direct each tanker to the proper hydrant to use either filling method most advantageously.

The most *inefficient* way to load a tanker is from the top. This is done when there are no direct fill inlets, and the tank cannot be filled at an acceptable rate through the pump and its associated piping. You fill through the hatch on the top with a hoseline, and a device similar to those used for pumping into portable tanks to reduce the reaction to the flow. Remember two things:

Figure 11-5. Gated wye on pumper outlet of hydrant provides better control with maximum flow.

Figure 11-6. Filling a tanker over the top.

1. This is extremely dangerous. Water, certain to overflow at times during filling, makes the metal on the tank very slippery. As hose handlers climb up on the tank before filling, and descend afterwards, they can fall and be seriously injured.

2. You lose too much time when you fill from the top because hose handlers must climb up, pull the hose up, insert it into the tank, and call for water. As water nears the top of the tank, it lowers the average fill rate. After the fill line is shut down, you have to remove it and close the hatch; then the hose crew has to climb down before the tanker can leave. All this activity keeps a tanker at the fill site twice as long as it would if a direct connection from the hydrant to the tank were used.

Given the ease of installing a direct fill connection on a tanker, there is no reason to operate a tanker that must be top-filled.

If you connect either 2½″- or 3″-lines directly from the hydrant to the tanker, the residual pressure and friction loss in the fill lines, and the available hydrant flow limit the maximum fill rate. To get a reasonable fill rate with small supply lines, you generally need to use dual lines. Some departments carry short sections of 3″-line to reduce friction loss when they fill directly from a hydrant. The lower friction loss allows tankers to fill faster, and increases the flow from the shuttle. Keep direct fill lines for tanker loading as short as hydrant distance permits.

When you set up a fill site from a hydrant, the goal is to connect enough fill lines, and control them individually to make connections to two tankers at the same time. Even though it is not a good practice to load both simultaneously, you can reduce time at the fill site if incoming tankers can be in place, with all their fill lines connected, before the tanker already loading is full. Tanker transitions are speedy and average fill rates are maintained at maximum practical levels. Since most hydrants have only two 2½″-small outlets, and you need dual lines to fill tankers with 2½ or 3″-hose, additional fittings are needed. Figures 11-7 and 11-8 show two different ways to set up dual loading stations from a single hydrant without using a pumper for filling.

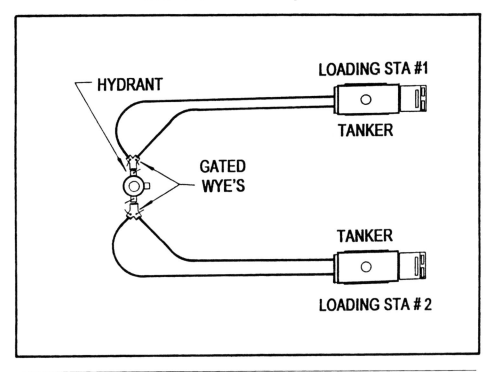

Figure 11-7. Loading two tankers directly from the 2½"-outlets of a hydrant.

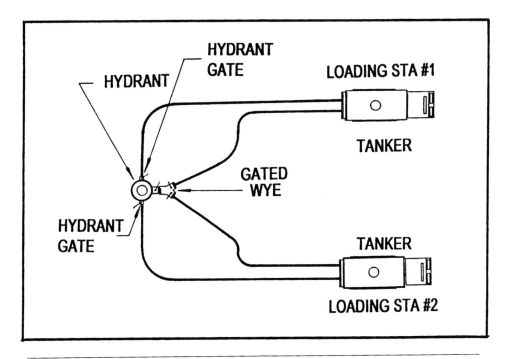

Figure 11-8. Using all outlets of a hydrant for loading two tankers.

LARGE DIAMETER HOSE (LDH) FOR FILLING

Large diameter hose is effective for loading tankers. Four- or 5''-LDH can replace a 4½''-soft suction hose for filling tankers through the dump valve if you have the necessary fittings and adapters. On some tankers there is a fill opening that lets you connect LDH directly to the tank. A standard 100-foot section of LDH offers you more flexibility for tanker positioning than does a 15-foot section of soft suction hose, and lets you take advantage of the high flow rate from the hydrant's large outlet.

You can fill tankers using large diameter hose, even tankers that have no way to connect directly to the tank. Connect the LDH to the large outlet on the hydrant and use it to supply a manifold or gated wye some distance from the hydrant where you have room for a good fill site operation. With the negligible friction loss in a short lay of LDH, multiple 2½''- or 3''-fill lines can be connected to the manifold. Each will have nearly the flow capacity of a direct connection to the hydrant.

Figure 11-9. Using LDH to load tankers by direct supply 75' from the hydrant.

FILLING WITH PUMPER CONNECTED TO THE HYDRANT

One advantage of the direct fill method is that there is no need for a pumper on the hydrant. If a tanker carries the needed fire hose, fittings, and adapters, the driver can set up a fill site, and fill the tanker from a hydrant, without waiting for help. However, not many tankers carry enough equipment to do this.

A better fill site option is to station a pumper at the hydrant, and connect all fill lines to the pump's discharge outlets. This method has four advantages:

1. Most pumpers have at least four 2½"-, or larger, discharge outlets, which can supply two loading stations, each with dual fill lines.
2. Pumpers generally have individual line gauges for each discharge outlet which tell the Fill Site Officer what pressure is being applied to each fill line. He or she can make sure the tanker is loaded as quickly as possible with no damage to the tank.

Figure 11-10. Water supply pumper loading tankers from a hydrant.

3. A pumper can increase the flow pressure from the hydrant enough to overcome the friction loss in the fill lines and the intake fitting of the tank and maintain the maximum flow rate. With a pumper on the hydrant, you can increase the average fill rate through a single line of 3''-hose from 500 GPM to 1,000 GPM.
4. Pumpers normally carry a full load of fire hose, plus the assorted appliances and adapters to make setup changes without calling for additional equipment.

When an empty tanker arrives at the fill site hydrant before the water supply pumper, consider connecting at least one fill line to the hydrant and filling the tanker. Let it return to the dump site with a full load of water without waiting for the pumper and leave the fill lines in place for other tankers to use until the engine company arrives and sets up a standard fill site. Even when a tanker has already used a hydrant for filling, and direct fill lines are in place, designate a pumper to go to the hydrant and set up a fill site as soon as possible.

ACCESSIBILITY

When you select a hydrant for filling tankers, remember that they generally have to turn around to return to the dump site, so you'll need several loading stations, spaced far enough apart to let them position and load without blocking arriving and departing tankers. For this you need the type of space found at shopping centers, warehouses, schools, parks, or other facilities. If the lot has an accessible fire hydrant, it makes an ideal location for a fill site. If no such lot is available, use a lightly traveled street, and if necessary, close it to make room for the fill site operation. Figure 11-11 shows one way to provide two loading stations in a side street. Keep traffic away from your personnel working around the apparatus. Wherever possible, isolate the fill site from all civilian traffic, or use traffic control officers to direct and keep the site safe.

DRAFTING TO FILL TANKERS

You should be able to fill tankers faster if a large body of water is available for drafting than if you use a typical rural hydrant system. When you select a draft site, consider whether:

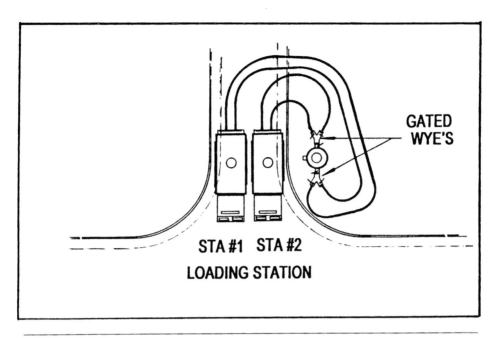

Figure 11-11. Setting up two loading stations at an intersection.

Figure 11-12. Using a parking lot for loading two tankers.

- It has enough water in storage, or sufficient flow, to fill tankers at a rate of 1,000 GPM or more, for the duration of the incident.
- The draft pumper can get close enough to reach the water with 20 feet of suction hose. If your department frequently must draft from a high lift, or you have trouble getting apparatus close to water sources, carry three sections of hard suction hose on your pumpers.
- The lift is ten feet or less from the eye of the impeller in the fire pump to the surface of the water at the draft site. You can use a pumper to fill tankers when the lift is more than 10 feet, but it will reduce your maximum flow rate by as much as 50% (See Figure 7-6, Chapter Seven: Suction Supply as a Water Source).
- The drafting pumper has the capacity to deliver water to the tanker as fast as the tank can accept it, or at a minimum rate of 1,000 GPM. You need a fill rate of 1,000 GPM to maintain a continuous flow of 750 GPM through a water shuttle. If you must use a 750-GPM or smaller pumper to fill tankers in the shuttle, you will have trouble transporting enough water to supply two 2½"-attack lines, much less a master stream.

Fill tankers with pumpers that are well maintained and in good condition. Although a typical fire department pumper rarely has to supply its rated capacity at a fire, when it is used to fill tankers, it should operate at or above its rated maximum flow to keep tanker fill times as short as possible.

Water supply pumpers should carry all the specialized equipment needed for fill site operations. Of all this equipment, probably the most important single item is a floating strainer, because with it a pumper can supply its rated capacity from any reliable source, even when water depth is less than 18 inches. Using a floating strainer also increases the supply's dependability, because as water is drawn into the strainer from the bottom, debris and vegetation floating on the water's surface are less likely to block the strainer and reduce or interrupt flow.

LOADING STATIONS

Water supply pumpers also should carry an assortment of fittings—clappered siamese, gated wyes, manifolds for LDH, quick-connect couplings, etc.—that can be used to set up loading stations for tanker filling. Two loading stations should be designated for each pumper at the fill site, and each loading station should have two 3"-fill lines,

Figure 11-13. Petroleum-type, quick-connect couplings.

Figure 11-14. Filling a tanker with a direct line from a water supply pumper at draft.

a single line of LDH, or both, to take advantage of the pumper's full rated capacity when filling tankers.

Water shuttles nearly always involve using water supply apparatus from more than one fire department, and in some areas, it is possible that not all tankers will have compatible hose thread connections. One solution is to use dedicated fill lines, with proper couplings, for each tanker or group of tankers. Or, you can use adapters to make the connections compatible, but you must provide adapters at each loading station and mark them for the appropriate tanker. Unfortunately, either solution results in lost time, and creates confusion at the fill site. Consider instead using quick-connect or quarter-turn Storz couplings on all fill lines and on each tanker to increase the fill site's efficiency and reliability. You can connect faster with these than with traditional hose couplings, and use any fill line with any tanker. This reduces the time lost in making and breaking connections. One fire department carries six sets of quick-connect adapters on the pumper generally used for filling tankers. These adapters are provided to tankers in a water shuttle that have fill inlets without them.

Generally, the closer your tanker loading station is to the pumper, the more efficient your operation will be. When the fill pumper is operating at draft within 100 feet of the tanker loading point, you can connect the fill lines directly to the pumper's discharge outlets and let the pump operator control it. With this layout, the pumper needs at least four 2½''-discharge outlets. If you draft with a 750-GPM pumper that has only three outlets, a gated 2½'' × 2½'' × 2½'' wye can be used as the fourth outlet. When your shuttle tankers are set up to use LDH for filling, a single 4- or 5''-hose can replace the dual 3''-lines at one of the loading stations. If the fill line is shorter than 200 feet, one line of 4''-LDH can transport the pumper's full capacity to the tanker. With 5''-LDH, the distance from the pumper to the loading station can be as much as 1,000 feet without a significant reduction in flow. Using a LDH fill line is practical only when all tankers that will use the loading station can connect to it.

If pumper-to-tanker fill lines are too long, friction loss limits maximum fill rate. However, drafting is often done in relatively inaccessible locations and tankers might lose more time getting close to the pumper to keep fill lines short and the flow rate high than could be saved by the higher fill rate.

Using LDH at fill sites gives high fill rates and allows tankers to load in an accessible location, some distance from the pumper that is operating from draft. With 5''-hose, you can fill tankers at a rate of 1,500 GPM, or more, at a loading station as far as 1,000 feet from the actual drafting site.

Figure 11-15. Five-inch LDH connected to direct tank fill line.

REMOTE LOADING STATIONS

If loading stations are set up in a remote location (some distance from the pumper supplying water from draft), fill lines should be controlled at the loading station. You can do this best by using a second fill pumper at the loading station, especially if the distance from the supply pumper is more than a few hundred feet. Then, supply the fill pumper from the water supply pumper through a conventional or open relay. This way the supply pumper and any relay pumpers can compensate for friction loss in the supply line and maintain a maximum flow rate while filling tankers. The advantage of an open relay versus a conventional closed relay is that an open relay has to supply only enough water to maintain the average flow from the shuttle; often, this is significantly less than the needed fill rate. When it drafts from a portable tank, the fill pumper draws from the tank's reserve supply to supplement the flow from the relay, and fills tankers at a higher rate. You can control each fill line from the fill pumper.

You do not need a second pumper with shorter distances because each loading station can be supplied by a hoseline from the supply pumper, with a gated wye at the fill point. Each fill line to the tanker is controlled at the loading station, making coordination simpler. Figure 11-16 shows one way to set up a remote fill site operation. If

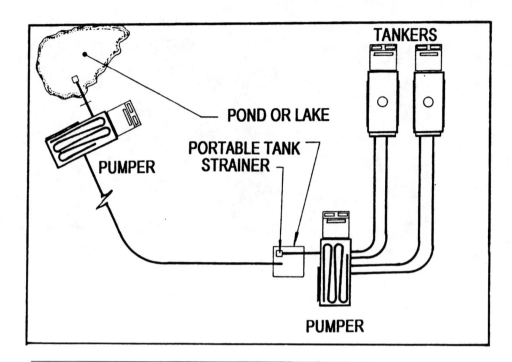

Figure 11-16. An open relay from a drafting pumper to a supply pumper for filling tankers.

your supply pumper is more than 100 feet from the remote loading station, use dual lines of 2½" or 3"-hose or a single line of LDH. If you have LDH, use a manifold that has enough gated outlets to connect all four fill lines to supply two loading stations. That way one person can control all the fill lines for both loading stations and make transitions faster after each tanker is loaded.

ORGANIZATION

Fill site organization is similar to dump site organization, but with important differences. Fill Site Officers generally work in isolated locations and can expect minimal help from incident commanders or other officers. Isolated, the Fill Site Officer lacks information and his or her crew often is unaware of events at the incident scene. If they cannot plan ahead, they cannot meet the demands of a changing situation. What helps is to establish and maintain reliable radio communication between the WSO and the fill site.

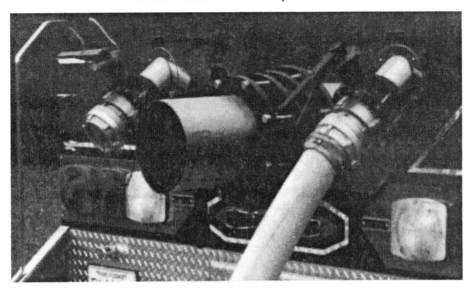

Figure 11-17. Two loading stations supplied by one LDH supply line, controlled at the loading point.

Figure 11-18. Four-inch LDH line supplying two loading stations with dual 3''-fill lines and one with a 4''-LDH fill line.

A typical fill site operation is complex. It requires a dedicated, well-trained, and organized team. Given its physical isolation from the incident command structure, it is important to delineate a fill site chain of command. A written standard operating procedure should be developed and followed. Below are specific responsibilities for the required members of a fill site team.

Fill Site Officer (FSO)

- Directs (and is responsible for) the fill site operation.
- Assigns jobs to all team members, and clarifies their responsibilities.
- Designs fill site (establishes supply, designates tanker loading stations, specifies fill lines, and arranges installation of valves, adapters, and special fittings).
- Arranges traffic patterns to avoid tanker delays and to minimize congestion and confusion.

Pump Operator

- Furnishes sufficient water to fill tankers as rapidly and safely as possible. (Even if tankers are filled directly from a hydrant, one team member must assume this responsibility and control the valves at the hydrant.)

If pumper is connected to a hydrant:
- Determines needed pressure to be supplied to each fill line.
- Sets pump to maintain pressure under rapidly changing conditions.
- Observes residual pressure when fill lines are opened so flow rate does not exceed hydrant capacity, and cause a vacuum in the water system. (When a tanker has been filled at a high flow rate, valves must be closed slowly to prevent water hammer.)

When filling from draft:
- Primes pump and maintains proper pressure to fill lines.
- Directs water flow back into source to avoid overheating pump and losing prime while awaiting tankers. (To fill tankers as rapidly as possible, pumper must operate at, or near, rated capacity with rapidly changing flow rates. Only the most skilled pump operators can handle this responsibility successfully.)

Figure 11-19. Booster line discharging into source to keep pump from overheating or losing prime when no tankers are loading.

Hose Handlers

- Connect and break fill line connections to tankers. (Hose handlers should be ready when incoming tanker approaches loading station, and make connection as soon as tanker wheels stop turning.)

When remote loading stations are being used:
- Open and close valves to the fill lines. (Use hose clamp to control flow if not enough valves available for each fill line.)
- Disconnect fill lines when tanker is filled.
- Signal driver that lines are broken and tanker is free to leave for dump site. (To prevent damage to equipment and injury to personnel, driver should never move vehicle until signal is received from hose handlers.)

Vent Crew

The responsibilities of the vent crew are more important during tanker filling than during tanker dumping. While inadequate venting may reduce the discharge rate in dumping, filling under pressure can cause damage to the tank if it is not adequately vented. Pressure will build up inside the tank if the air cannot escape as fast as water enters. Except in specialized applications, the tank on a fire department tanker is not a pressure vessel; it can be damaged if it is pressurized. Properly constructed tankers have adequate venting. The first time a tanker arrives at the fill site in a shuttle, the Fill Site Officer should check to see whether the venting is adequate. When the hatch is unlatched, it should remain closed by gravity. If the vent area is not large enough to prevent a pressure buildup, the hatch cover will be forced open, and air will be expelled under pressure at a high velocity. If you have any doubts, leave the hatch cover unlatched while the tanker moves back and forth in the shuttle. If pressure builds up during filling, the hatch will open and relieve it.

If the tanker can be filled only from the top, the vent crew handles the fill lines. All lines should be tied down before they are charged, and special filling devices must be used to reduce the reaction when water is discharged from the fill line.

Traffic Control Officer (FSO or others)

- Directs tanker drivers to loading stations.
- Protects point of entrance to and from highway. (Fill site team responsible for safe and efficient movement of vehicles at fill site.)

Safety Officer Concerns

- Tankers are filling at extremely high rates of flow with fill lines operating at high pressure.
- Fill lines have to be brought into position to fill at one loading station while activity is taking place at another loading station, often in close proximity to each other.
- When operating from draft, there is the danger of personnel falling into the water.
- Fill sites often involve using a hydrant located along a busy street, with danger from civilian vehicles.

Whenever possible, a specific individual should be appointed safety officer and assigned the responsibility of observing the overall fill site operation. The safety officer should watch for unsafe conditions or operating practices that present a danger to personnel or equipment. The fill site officer can take on this added responsibility, but a dedicated safety officer, one who is not emotionally involved in the operation, will provide an added margin of safety.

FILL SITE DESIGN

As soon as you realize that you need a water shuttle, assign a responding engine company to go to a specific water supply point and set up a tanker fill site. If you can, include an extra engine company on the initial alarm whose sole responsibility is water supply. During your planning process, identify and mark water supply points you can use to fill tankers. Alert all engine companies and mutual aid departments to the water supply points in the area, and give dispatch detailed information about, and directions to, water supply points.

After you select the water supply point, have the WSO determine which engine company can best set up a fill site. If you diverted the engine company en route, it should be ready to fill tankers five minutes after arriving. After you determine travel distance, and the amount of equipment responding, decide whether to request an additional engine company for this assignment. Remember: there is no time to lose! If you are using a hydrant for filling, have the fill site crew, or at least the FSO, come from the fire department responsible for the area where the hydrant is located. An officer from the department with jurisdiction knows more about the system's capacity, its limitations, and the measures to take if problems arise.

Position the fill pumper to provide a safe working location for the pump operator, with minimum interference with tanker movement. If possible, locate loading stations in the pump operator's line of vision. Then the pump operator can remove pressure from fill lines promptly with minimal damage, and less chance of injury to personnel.

Plan loading stations for tanker maneuverability and to minimize delays. Arrange the fill site so tankers can do their maneuvering before they are loaded, when they are lighter and easier to handle. Direct the heaviest tankers (those equipped with tandem axles or tractor trailer units) to the loading position that requires the least maneuvering, and let smaller tankers load where there is less accessibility. Consider a staging area to simplify traffic patterns and keep tanker waiting time to a minimum.

Arrange fill lines so that the flow control person can see the entire hoseline and the tanker connection point. Move the control point out of traffic flow where moving tankers are not a danger. Reduce confusion by carrying different-colored tapes or ribbons to attach to both ends of a fill line. Refer to fill lines by color; hose handlers and the firefighter who controls the valves will all know which line you want pressurized or shut down.

Fill lines should be able to handle the fill pumper's maximum flow. As a general rule, have two 3''-lines available at both loading stations. With LDH, a single line is adequate. The fill site must accommodate tankers that load from LDH and tankers that use 3''-fill lines. You can do this by setting up one loading station with a single line of LDH, and one with dual 3''-fill lines. Or, arrange each loading station for two 3''-lines and one LDH line, individually controlled. This is enough flexibility to load any tanker at either station.

When you cannot see incoming and outgoing tankers from the fill site, notify the traffic control officer so that he or she can assign someone to protect the intersection. Tankers should be directed by the traffic control person until they are clear of the road and at the fill site.

FILLING TANKERS

When each tanker arrives at the fill site for the first time, the fill site officer and hose handlers determine how to best fill it. If a high flow inlet to the tank permits filling at the maximum rate with only one line, you can save time by using it rather than connecting two lines each

Figure 11-20. Fill site for loading tankers directly from supply tank.

time. If only 2½″-hose is available for fill lines but the source can supply more than 500 GPM, you will need two fill lines. A single line of 3″-hose can handle as much as 1,000 GPM. Higher fill rates require dual lines, or a single line of LDH.

If you can connect only a single fill line to a tanker, or if you have no direct tank fill connection, you might have to fill from the top. This, the least effective way to fill tankers, is a last resort unless a special loading arrangement like the one shown in Figure 11-20 has been provided.

When you first fill a tanker, check the venting capabilities. Generally, this means that someone has to get on top of the tanker. Observe the vent, and check the velocity of the air as it is forced out of the tank to determine whether the tank is vented adequately. If not, open the hatch to prevent pressure buildup each time the tanker is filled. **This is a dangerous operation:** no one should climb up on the vehicle until it stops. After you load it, do not let the driver move the vehicle until the fill site officer signals that everyone is clear of the tank and it is safe to leave. Make filling safer by designating the person who goes on top of the tanker as the traffic control person. Do not move the vehicle until the All-Clear signal is given. When a tanker is vented adequately, ignore this aspect of filling.

PUMP OPERATION AND FLOW CONTROL

The fill site team's primary objective is to load tankers and get them on the road as quickly as possible. How quickly? With a well-designed fill site, your team should have tankers returning to the dump site two minutes after they arrive at the fill site. This requires a high flow rate. Do not lose time making the transition to filling the next tanker as each one is loaded.

ACHIEVING A MAXIMUM FILL RATE

Using direct fill lines from a hydrant gives the fill site crew few options to increase the maximum fill rate. Only the hydrant's residual pressure is available to overcome the friction loss in the fill lines; you will need LDH or dual lines to fill the average tanker in under two minutes. With direct fill lines, the shorter the line, and the fewer appliances used, the higher the fill rate will be.

When you use a pumper to fill, either from a hydrant or from draft, it can supply the pressure needed to fill the tanker quickly. The most important aspects of operating the pumper are to recognize when the maximum fill rate has been reached, and sustain it until the tanker is filled, water is coming out of the overflow.

If a hydrant supplies the fill pumper, the compound gauge on the intake is the best indication that the maximum fill rate has been reached. For safety and to prevent damage to the system, keep the hydrant's residual pressure to least 20 PSI. When the valve opens and the tanker begins to fill, advance the throttle on the pumper until the residual pressure on the intake reaches 20 PSI or until the maximum fill rate specified for the tanker being filled is reached. Then you have reached the maximum flow that can safely be taken from the water system; adjust the throttle to keep it at 20 PSI or the maximum flow rate until the tanker is filled. If the discharge pressure reaches 150 PSI or the maximum fill rate before the residual pressure reaches 20 PSI, it means that the system can supply water faster than the tanker will accept it. To prevent damage to the tanker, do not let discharge pressures exceed manufacturers' specifications.

If you are operating the fill pumper from draft, the vacuum reading on the compound gauge increases when the valve opens and water flows into the tanker. Increase the throttle setting and the rate of flow goes up, friction loss in the suction hose is greater, and the vacuum reading approaches 20 inches. Increase the throttle while you fill tankers until the pump goes into cavitation or the maximum fill rate of the MWS apparatus is reached. The point of cavitation is when the flow reaches a point where further increases in the throttle setting do not change the discharge pressure. You then decrease the throttle until the pressure begins to drop. At that point, the pumper is delivering all the water it can move and is filling the tanker at the maximum possible flow rate.

Do not let pump operators use any more pressure than necessary to achieve the maximum fill rate. With relatively unrestricted tank fill connections (the case when you fill through a dump valve or a direct fill line), approximately 20 PSI is needed at the end of the supply line. If you must fill through the pump on the tanker, you will need slightly more pressure, but the discharge pressure from the fill line is, to a certain extent, self-regulating. An increase in discharge pressure at the fill pumper causes a higher rate of flow into the tank. The higher flow causes an increase in friction loss in the fill line, which limits the actual increase in pressure at the tank inlet.

Use the following formula to calculate the needed discharge pressure at the fill pumper

$$Ep = FL \pm HP + 20$$

where

- Ep = discharge pressure needed to develop a specified rate of flow for a fill site
- FL = friction loss anticipated in the fill lines at the specified rate of flow (Allow 15 PSI for each wye or manifold in addition to the friction loss in the hoseline.)
- HP = amount of pressure required to overcome the head pressure caused by differences in elevation between the tanker loading station and the fill pumper
- 20 = pressure needed to create the desired flow rate into the tank

For example, a fill site is set up to fill tankers at a rate of 1,000 GPM through a 3"-fill line, 100 feet long. The fill line is controlled from the pump operator's panel on the fill pumper, with no difference in elevation between the fill pumper and the loading point. The expected friction loss in 100 feet of 3"-line flowing 1,000 GPM would be approximately 100 PSI. The discharge pressure needed at the fill pumper would be 120 PSI as follows:

$$100 \text{ (FL)} \pm 0 \text{ (HP)} + 20 = 120 \text{ PSI}$$

If you modified this loading station to provide a 100-foot section of 3"-hose feeding a 2½" × 2½" × 2½" gated wye, and two 100-foot sections of 3"-hose to the tank fill connections, 160 PSI would be needed. The gated wye would add 15 PSI friction loss, and the 100-foot fill lines would add 25 PSI friction loss since the flow would divide equally between the two lines. The needed engine pressure would increase from 120 PSI to 160 PSI as follows:

$$EP = 100 + 15 + 25 \text{ (FL)} + 0 \text{ (HP)} + 20 = 160 \text{ PSI}$$

These calculations show that a single line of 3"-hose could supply 1,000 GPM if the distance from the pumper to the tanker were 100 feet or less and if the pump were capable of supplying it with a discharge pressure of 120 PSI. For distances longer than 100 feet, you would need dual lines or LDH to fill at a rate of 1,000 GPM. With only 2½"-hose for filling, you would need multiple hoselines to avoid a reduced fill rate.

FILL RATE LIMITATIONS

According to NFPA 1903, *Standard for Mobile Water Supplies* (1991), a tanker must accept water at a rate of 1,000 GPM without damaging the tank. Unless you order a higher flow rate, the manufacturer generally will not guarantee the tank against damage incurred because of higher fill rates. Some tanks are labeled near the fill inlet, warning against fill rates higher than 1,000 GPM. Unless the pumper is equipped with flow meters, it is impossible for the pump operator and fill site officer to determine the flow rate when filling. If you are using a 1,250- or 1,500-GPM pumper to fill, you can easily exceed 1,000 GPM with adequate fill lines. From an efficiency standpoint, tankers that carry more than 2,000 gallons should be able to be filled at a higher rate.

Some manufacturers are building tankers that can be filled at a rate of 2,000 GPM or more without damaging the tank, but these companies generally are smaller and specialize in mobile water supply apparatus. Although most major manufacturers are reluctant to build tanks that they can guarantee at flow rates that exceed 1,000 GPM, it is cost effective to pay more for this capability, because not only can each tanker deliver more water, but you get a safety margin and comfort level knowing that the tank will not be damaged if the fill rate inadvertently surpasses 1,000 GPM during a high-capacity water shuttle.

CONTROLLING FLOW

To minimize both handling time and time spent at the fill site, open the valves that control the flow to the standby tanker immediately after the tanker being filled overflows. Opening and closing these valves should be a push-pull operation, that is, the fill lines to the loaded tanker are shut down when the fill lines to the empty one are opened. A delay of 30 seconds in starting to fill the next tanker decreases the average fill rate by as much as 20%, even though the flow rate remains the same. This operation is smoother if one person can control all of the valves.

If the fill inlet at the tanker has a shutoff valve, do not use it to stop the flow when the tanker overflows. Cut the fill lines off at the supply end rather than at the tanker. If you close the valve at the tanker before the valve at the supply end, pressure gets trapped in the hoseline, and it is difficult to break the connection while the fill line is pressurized.

If the coupling is disconnected, pressure can cause the end of the line to whip in a way that injures personnel or damages equipment. If you close the valve at the supply end first, pressure in the line is relieved into the tank before the tank fill valve closes. If tankers are equipped with check valves on the tank inlet instead of a positive acting valve, this will not happen. You can install a check valve externally to simplify the operation by connecting a clappered siamese to the direct fill inlet. The clappered siamese serves as a check valve after filling the tank, and you can leave the gate valve at the inlet open during the shuttle.

COORDINATION

Coordination is the key to minimizing time at the fill site. The fill site officer can save time and maintain close control of every aspect of the operation by:

1. Directing tanker drivers to the proper loading station without any delay or uncertainty;
2. Directing each tanker to fill at the loading station where it can best be loaded. Consider line sizes, correct fittings, and accessibility when you select the best loading station;
3. Removing lines without delay as tankers are filled, and directing drivers to leave the fill site when all personnel are clear. The person who directs the drivers should be in their line of vision at all times; and
4. Remaining in constant contact with the WSO and the dump site officer and keeping them informed of any difficulties that may be experienced.

12. High-Capacity Water Shuttle

WHAT IS A HIGH-CAPACITY WATER SHUTTLE?

UNLIKE A RELATIVELY EASY-to-maintain 500 GPM fire flow water shuttle, a High-Capacity Water Shuttle is a shuttle with a design flow of 1,000 GPM or more that requires a dump site where you can store more water (at least 3,000 gallons of usable water) to maintain a consistent flow of 1,000 GPM. Greater flows require proportionately higher amounts of storage, and typically use multiple portable tanks, a large tanker for use as a nurse, or some combination of the two. Six or more tankers are needed to shuttle water, and routing and traffic patterns have to be planned to minimize congestion. In addition to a good hydrant system, high-capacity pumpers, and multiple locations for filling tankers, a fire department needs specialized techniques and operating procedures to operate a high-capacity water shuttle successfully. Most rural and suburban fire departments have neither.

Most rural and suburban fire departments have areas of concentrated development or large buildings in areas not served by hydrants that would require fire flows of 1,000 GPM or more. One way they can support heavy fire streams is to develop the capability to operate a high-capacity water shuttle.

DUMP SITE OPERATION

Maintaining a dependable water supply by using a water shuttle requires that you select a dump site where you can store at least enough water to operate the water supply pumper for three minutes without a tanker unloading. This means that the usable water in storage must be at least three times the maximum Needed Fire Flow (NFF). For a high-capacity water shuttle you need to store at least 3,000 gallons of usable water at the dump site to provide the required reserve.

The amount of usable water in a portable tank or nurse tanker will always be somewhat less than its total capacity. The variables that affect the percentage of the water stored in the tank that is readily available for use are outlined below.

Usable Water—Portable Tank

The terrain, positioning of the tanks, and the way the pumper is set up to draft all have an effect on the amount of usable water available from a portable tank. As a general rule, consider the amount of usable water to be no more than 75% of a tank's maximum capacity. While you sometimes can arrange a dump site to improve the percentage of usable water, it usually is better to err on the side of more than to come up short of the needed reserve.

Usable Water—Nurse Tanker

A tanker used as a nurse typically provides less usable water than the same tanker used as a shuttle tanker. In a high-capacity water shuttle, unless a tanker pumper with a 1,000-GPM pump is used as a nurse, the water supply pumper must be able to draft 1,000 GPM or more from the nurse tanker instead of depending on the tanker's pump to supply the attack pumper. To do this, a hard suction hose, at least five inches in diameter, must be connected to an outlet from the tank. The large outlets designed for gravity dumping from a tanker may be less than ideal for drafting.

Dump valves frequently are installed in the back wall of the tank. When you are in a dumping mode, water continues to flow until the tank is empty. However, when a pumper is drafting from the tank, the flow stops while there is still a significant quantity of water in the tank, and when the water level drops below the top of the dump opening, the pump can draw air and lose its prime. With a six-inch dump valve, there will be a minimum of six inches of water left in the tank at this point. The whirlpool effect can cause the pump to lose its prime even sooner unless some type of anti-swirl device has been installed inside the tank. Depending on the height of the tank, you could have as much as 20% of your water left in the tank when your water supply pumper is unable to continue to provide the NFF.

Terrain is a factor in the amount of unusable water in a nurse tanker as well as in a portable tank because if a tank is not level when it is being filled, water begins to discharge from the overflow before it is full and you will not be able to use the tank's full capacity. If the vent is not at the tank's high point as it is positioned, air pockets can form, limiting the amount of water that can be stored. If the pumper is drafting from the high end of the tank, water trapped in the low end of the tank when the pump loses its prime will not be available for use.

Baffles that are too restrictive also can reduce the percentage of usable water in a nurse tanker because they limit how fast water transfers between compartments. Experience and testing are the best indicators to use with a particular nurse tanker. So, a tanker designed for a 600-GPM dump rate may not be able to maintain a 1,000-GPM flow.

The rule of thumb for estimating the amount of usable water in a portable tank is a good guideline for nurse tankers as well: the usable water in a nurse tanker should be estimated to be no more than 90% of its maximum capacity when using the pump on the tanker, or 75% when an external pumper is drafting from the tank, **unless you have any indications to the contrary**.

MULTI-TANK OPERATION

Portable tanks larger than 3,000 gallons are rarely carried on tankers because such large tanks are unwieldy, hard to handle, and require too much space to set up and use. If you apply the 75% factor to the 3,000-gallon maximum capacity, you would need more than one tank any time the desired fire flow exceeds 750 GPM, so for practical reasons plan to use more than one portable tank whenever your water shuttle requires more than 500 GPM.

This applies to nurse tankers as well. Most fire department tankers carry less than 2,500 gallons, adequate storage for flows up to 700 GPM. A trailer-type tanker that carries 6,000 gallons might have the reserve to support a 1,250-GPM shuttle, but only if you can pump from it at this rate. Given the limited number of intake connections for filling on most tankers, and the difficulty in drafting from a nurse tanker at rates greater than 500 GPM, it still seems like a good idea to plan for multiple tanks when more than 500 GPM is needed.

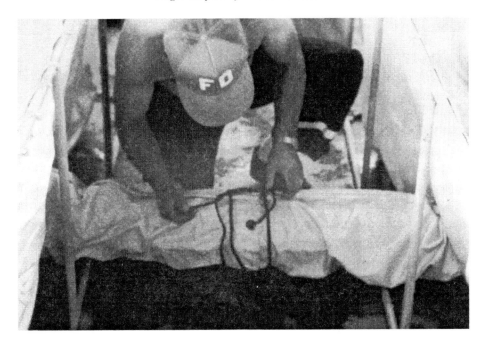

Figure 12-1. Connecting drain tubes using a section of 6″-PVC pipe.

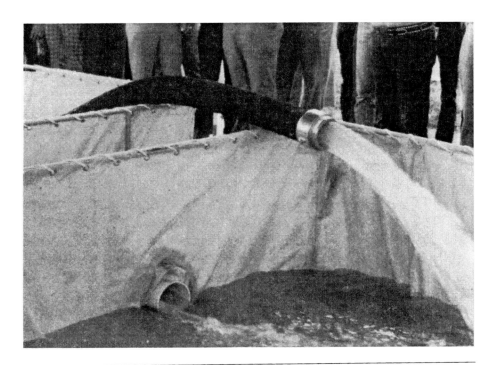

Figure 12-2. Drain tube connection yields a very low rate of flow compared to the power transfer through the suction hose.

Water Transfer

Remember that even if you are using multiple portable tanks, the water supply pumper generally can draft only from one of them; your other tanks are there to create a reserve supply to maintain the level in the supply tank when it gets too low. If the water supply pumper has more than one gated intake, it is possible to use a hard suction hose in each one; this enables the pumper to draft from either of two tanks. Once you prime the pump and evacuate the air from both suction hoses, the pumper can draft from either tank by opening and closing the intake valves. Not all pumpers can do this. It is more practical to use a transfer device to move water into the supply tank as needed.

Several different methods have been developed to transfer water between portable tanks, including connecting the tanks together through the drain openings using gravity transfer devices which have a siphon action to maintain flow, and using a 1½"-hose equipped with a transfer device to drive the water through a section of hard suction hose or other conduit between tanks.

Any method that depends on the head pressure created by differences in water depth in adjacent tanks to equalize the level in the tanks, has three basic problems:

- First, the flow rate is not high enough to maintain a high-capacity water shuttle. When the difference is greatest, i.e., one tank full and the other empty, a maximum pressure differential of 1 PSI would be created between tanks and would move only a limited amount of water. This flow rate would decrease steadily as the level dropped in the full tank and increased in the other.
- Second, water will flow in either direction through these devices, so if the level in the supply tank is higher than in the reserve tank, water would drain away from the supply tank. Then the level in both tanks would equalize, making it difficult to maintain an adequate water supply for the pumper to draft.
- Third, there is no way to control the transfer of water! Once the siphon action begins or the tanks are connected, the level in interconnected tanks will equalize until the siphon is broken, or the connecting line is closed off.

The solution? A power transfer device.

A power transfer device addresses all three of the aforementioned problems. It works this way: a 1½"-hoseline discharging into the end of a hard suction hose or other conduit that is submerged in the reserve tank transfers water into the supply tank by a Venturi-type action.

Figure 12-3. Homemade power transfer device using solid stream nozzle from standpipe hose.

Experience has shown this transfer method to be efficient. The maximum flow rate is limited primarily by the size of the transfer tube, but a well-designed device should be able to transfer water at the following rates:

                4½″ ----------------------------450 GPM
                5″ -----------------------------550 GPM
                6″ -----------------------------650 GPM

These rates are based on using a ½″-solid bore nozzle on the 1½″-hose with a discharge pressure of 100 PSI on the pumper that supplies it. Note that in addition to the estimated flow through the suction hose, approximately 50–150 GPM would be flowing from the 1½″-transfer line. If you are using the water supply pumper to operate the transfer device, this flow comes from the supply tank and is just circulating. To find out how much water is actually being transferred into the supply tank from the reserve tank, you need to subtract the flow from the transfer line from the total flow through the suction hose or transfer tube.

Another reason to use a power transfer device is that the flow can be controlled by the pump operator, either on the supply pumper or one being used to supply pressure to the transfer device. As long as the discharge end of the suction hose or pipe used for transferring is not under water, flow will stop when you close the valve to the transfer line, and begin when you open it. The transfer rate is controlled by the nozzle pressure on the 1½″-line and the flow rate can be controlled by partially closing the valve and reducing the pressure supplied to the transfer line.

The transfer flow is controlled by two variables: the velocity of the water from the power transfer line and the size of the transfer device. You can use a 1½″-line without a nozzle, but you would need an inordinate amount of water to accomplish the transfer. You can get a higher transfer rate with a lower flow in the 1½″-transfer line by installing a smooth bore nozzle, or an adapter using a flow device that will generate the maximum velocity of the water entering the transfer tube. A number of do-it-yourself devices have been used for this purpose and there is a full line of appliances available from different equipment manufacturers.

Figure 12-2 shows a do-it-yourself device connected to the suction hose being used to transfer water between tanks. Figure 12-4 shows another variation. Here the drains on two tanks have been connected with an 8-inch PVC pipe. A power transfer device driven by a 1½″-hoseline is used to control the movement of water. With the large waterway, this device can transfer close to 1,000 GPM, if needed.

One innovation is to insert a 1½″-nozzle into the waterway of a low-level strainer as shown in Figure 12-5. You can then use the low-level strainer to draft from a portable tank, or as a power transfer device connected to a section of hard suction hose. This works well for a

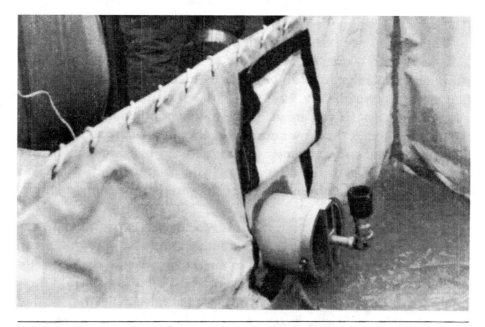

Figure 12-4. Homemade power transfer device for 8″-connection to drain tube.

Figure 12-5. Low-level strainer modified to use 1½"-hose for power transfer.

power transfer, but limits your flow if you are drafting from a portable tank. In a recent series of tests that used a 1,250-GPM pumper to draft through a 6"-hard suction hose, 950 GPM was the maximum flow available with the modified low-level strainer. A barrel strainer or a low-level strainer without the 1½"-fitting used with the same layout yielded an 1,150-GPM flow. It appeared that the nozzle inserted into the waterway restricted the flow due to increased friction loss and limited the pumper's ability to supply its rated capacity. Used in conjunction with an automatic suction valve on the pumper inlet, the maximum flow was reduced almost 400 GPM below the capacity of the pump with an unrestricted suction line.

These devices—automatic suction valves and low-level strainers with power transfer nozzles—are convenient to use and offer great flexibility to standard water shuttle operations. At the same time, they restrict flow, and are of limited value in high-capacity water shuttle operations.

Controlling water transfer using a separate pumper instead of the water supply pumper to provide the needed pressure is advantageous. Figure 12-6 is of a high-capacity water shuttle that is flowing 2,850 GPM from two water supply pumps. The third pumper is drafting from the reserve tank, and can transfer water into either of the tanks as needed.

Some of the advantages of using a transfer pumper are that:

Figure 12-6. High-capacity water shuttle flowing 2,850 GPM using separate pumper to transfer water between tanks.

1. A transfer pumper can set the discharge pressure to the transfer device to provide the needed flow rate since the water supply pumper is limited by the need to supply water to the attack pumper.
2. The transfer pumper can draft from the reserve tank. If the water level in the supply tank gets too low, the water supply pumper could lose its prime and be unable to transfer water from the reserve tank.
3. When you are drafting from the reserve tank, the water from the transfer line supplements the water being transferred through the suction hose.
4. By eliminating any other responsibilities, the pump operator can concentrate on water transfer with a minimum amount of distraction.
5. Water used to operate the power transfer device will not diminish the pump capacity of the water supply pumpers.

You can use a portable pump or a PTO pump on a brush tank to draft from a reserve tank and supply pressure to a power transfer device if no pumper is available.

The transfer is crucial to the success of a high-capacity water shuttle. When a quick-dump tanker approaches the dump site, allow the water level in the supply tank to drop enough to give the tanker a place to immediately dump its load. When a slower tanker (average dump rate less than the fire flow) is unloading, enough water must be transferred from the reserve tank at the same time to maintain a safe working level in the supply tank.

When you have multiple supply pumpers operating with a separate transfer pumper, the complexity of the dump site operation increases, and you should consider putting one person in charge of the transfer. Give this person a position with a vantage point, for example, a top-control pumper. At times, you may need to turn the device around and transfer water into the reserve tank from one of the supply tanks. This may happen when large tankers are dumping into a supply tank that cannot handle the water as fast as they can dump it. Only sound organization and timely instructions to all parties can provide the flexibility needed to make this type of operation work well.

COMBINED OPERATION

A combined nurse tanker and portable tank operation is another way to provide the reserve needed for a high-capacity water shuttle. If you use a tanker pumper for a nurse, it can draft from a portable tank or tanks and use the water in its tank to supplement water from draft. This gives your shuttle greater flexibility. When shuttle tankers are equipped with fire pumps of 500 GPM or more, they can pump directly into the nurse tanker while quick-dump and jet dump tankers can take advantage of their ability to unload quickly by dumping into the portable tank. At that point, the combination nurse tanker and water supply pumper can take water from either the portable tank or its own tank.

A nurse tanker can be used as a supplemental supply in conjunction with a portable tank. When a portable tank is used for the primary storage, the nurse tanker can maintain flow by discharging into the portable tank when the water drops to a dangerously low level. It also will provide additional storage for tankers to unload when the portable tank is full.

POSITIONING

Where you position portable tanks, nurse tankers and water supply pumpers is critical to the design of a high-capacity dump site. With the

Figure 12-7. Combined operation using a flexible tank, rigid tank, and nurse tanker for storage.

number of tankers involved, all tanks need to be accessible by each of the tankers. This keeps maneuvering time to a minimum.

High-capacity water shuttles often require more than one pumper to provide the NFF. Because pumpers larger than 1,250 GPM are rare in rural and suburban fire departments, more than one pumper would be needed to supply a higher flow rate. Using power transfer devices to move water between tanks also uses some of the fire pump's capacity. In general, when a water shuttle is designed to supply more than 1,000 GPM, it should include more than one water supply pumper. In some cases, the best solution is to operate two completely separate water shuttles with different dump and fill sites. On the other hand, using two pumpers at the same dump site provides more reliability and flexibility, since the water in the reserve tank, or tanks, can be used by either pumper as required.

Position a supply tank so that the water supply pumper can draft through a hard suction hose. The ideal position allows the strainer to take water from a corner of the tank where the turbulence caused by tanker dumping is minimized as illustrated in Figures 12-8 and 12-9.

Since a high-capacity water shuttle rarely starts out that way, the attack pumper, or initial water supply pumper, improperly positioned, can block access for additional apparatus and storage tanks. To avoid access problems have a water supply preplan for all target hazards, and

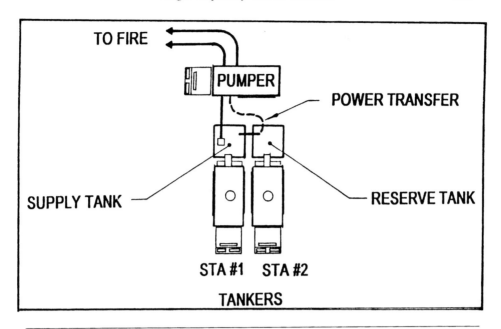

Figure 12-8. High-capacity water shuttle using two tanks and one water supply pumper.

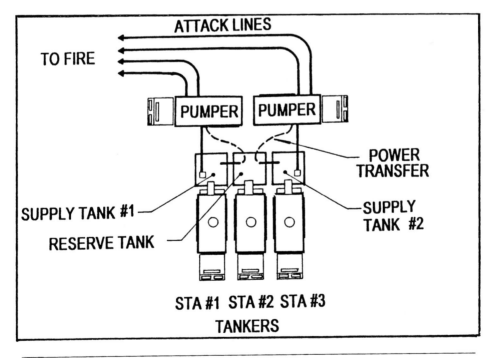

Figure 12-9. High-capacity water shuttle using three tanks and two water supply pumpers.

implement it any time an alarm is received. Design the plan so that the water supply can develop from an initial attack pumper and single storage tank to a high-capacity water shuttle in case the fire cannot be brought under control quickly.

One way to expand the water supply is to have the high-capacity water shuttle terminate in a location different from that of the initial water supply pumper. Water supply pumpers at a new dump site establish a relay that would supply enough water to the fireground to keep the initial folding tank or nurse tanker full.

Operator safety is the most important factor in setting up a high-capacity water shuttle dump site. With tankers maneuvering, backing into position at multiple tanks and all the accompanying activity, accidents can happen. One way to minimize the danger is to position tanks properly.

As additional tanks are set up and put into service, they must be located so that water can be transferred readily to the supply tank, where the water supply pumper is drafting. When the supply tank is full, tankers can dump into these reserve tanks. When more than one water supply pumper is in use, reserve tanks should be positioned so that water can be transferred into either of the supply tanks; to do this, the reserve tanks need to be located within two feet of the supply tanks. A good operating practice is to position the tanks at least 18 inches apart so that personnel have a safe place to work (without the danger of getting caught between a storage tank and an approaching tanker).

Figure 12-10. Three portable tanks arranged for working space between them.

Figure 12-11. Supplemental unloading station remote from dump site.

Figure 12-12. Pumper tanker at supplemental unloading station discharges into supply tank rather than reserve tank.

Figure 12-10 shows three tanks in a parallel position, with working room between each of them.

Among the different ways to position tanks are: staggered like a Christmas tree, located corner to corner in a diamond pattern, or positioned side by side. What is important is that the water shuttle design, and the position of the tanks, are adapted to the available space. Tanks should be positioned so that tankers can back into position easily to unload, and keep their maneuvering time to a minimum.

SUPPLEMENTAL DUMPING STATIONS

One way to avoid the congestion that can hamper smooth dump site operations is to minimize movement around storage tanks. This is possible if some of the tankers are able to pump off at a high flow rate. You can do this by establishing remote unloading stations far enough away from the tanks to prevent interference with dumping tankers. This way some of them can unload without blocking access to the tanks by those tankers that have to be positioned close enough to dump directly to take advantage of their best dump rate.

FILL SITE OPERATION

High-capacity shuttles need high-capacity fill sites. While this might seem to be an elementary observation, the need is sometimes

Figure 12-13. Remote fill site supplied by a pumper at draft through 500' of 5''-LDH filling tanker at a rate in excess of 2,000 GPM.

Figure 12-14. Two pumpers drafting from the same source to fill tankers.

overlooked. A 1,000-GPM pumper rarely can supply a 1,000-GPM water shuttle. It often is not possible to sustain an average flow rate that exceeds 75% of the peak flow from a fill site. There is also a limit to how many tankers can be loaded efficiently at one location. To minimize lost time, each pumper should supply at least two loading stations. Some fill sites can accommodate two pumpers at draft, and provide for as many as four loading stations. You can use large diameter hose to increase the distance between loading stations without seriously reducing the maximum flow rate. While all of these options are helpful, a high-capacity water shuttle generally needs additional fill sites. Establishing a second fill site any time more than 500 GPM is required from a water shuttle should be a standard practice. Even when you have high-capacity pumpers you still need an additional site each time the average flow from an existing location exceeds 500 GPM. This not only minimizes tanker congestion, but gives you greater reliability if problems develop at one of your fill sites.

Filling From a Hydrant

There is a natural inclination to fill from a hydrant when one is available. Doesn't it seem simple to go to a hydrant, connect a 2½"-line, and fill the tanker? However, depending on hydrant pressure to fill a single line of 2½" simply will not flow enough water to support a high-capacity water shuttle. High-capacity filling requires connecting a pumper to the hydrant with a supply line large enough to allow the full capacity of the hydrant to be used. In addition to the increased flow, connecting a pumper to the hydrant provides the outlets needed to supply additional fill lines. An adequate number of fill lines, with the proper hose threads or adapters already installed, must be ready and waiting each time a tanker arrives.

While it may be best to fill tankers directly from the hydrant initially, you need to detail an engine company as soon as possible to set up a high-capacity fill site with a pumper connected to the hydrant. When you preplan target hazards, include a reminder to fill tankers rapidly so that an adequate water supply can be maintained. Assign enough apparatus and personnel ahead of time to operate a high-capacity fill site efficiently.

DRAFTING SITES

You need a high-capacity pumper to maintain a high flow rate when you fill tankers from draft. Arrange a drafting site that will give you the greatest flow from the pumper you use.

Low-capacity water shuttle operations can incorporate mini-pumpers or portable pumps for filling, but these are not adequate for fill sites expected to achieve peak flows of more than 1,000 GPM. For a high-capacity fill site, use the largest available pumper and best drafting location. It is essential that the lift from the surface of the water to the pump be as low as possible, and definitely not more than 10 feet if you plan to use the pump's full capacity. In one exercise in New Jersey, a 1,250-GPM pumper filled tankers at a rate of 2,000 GPM. With a very low lift, two 6"-hard suction lines, and 5"-hose used to transport water to the fill site, an extremely high fill rate was achieved. When lift increases, maximum flow rate decreases. At 10 feet of lift, a pump can be expected to deliver its rated capacity, but as the lift gets higher, the capacity decreases. Do not expect a pump operating with a lift of 20 feet to supply a maximum flow that is more than 60% of its rated capacity.

Fill lines are important. To achieve peak flow, you may have to operate the fill site at a distance from the pumper supplying it, and, if so, the length and size of the fill lines may limit the fill rate, thereby increasing the time required to fill the tankers. Large diameter hose or multiple lines, however, can maintain a high flow rate. Maintaining a pressure of somewhere between 125 and 150 PSI to the fill line (50 or

100 feet of 2½″- or 3″-hose) at the loading station has been shown to achieve a maximum fill rate.

CAUTION: THIS CAN BE DONE ONLY IF THE TANKERS ARE ADEQUATELY VENTED TO HANDLE THE FLOW WITHOUT A DANGEROUS PRESSURE BUILDUP INSIDE THE TANK.

For the best fill rate, use hoselines between the pumper and the loading station that can handle the desired flow with a friction loss that will maintain the desired pressure to the fill lines.

TRAVEL TIME AND MANEUVERABILITY

Tanker routing is extremely important. While it is never desirable for tankers to pass each other, traffic patterns become even more important in multiple-tanker high-capacity water shuttles. When you need extremely high flow rates, even short delays can cause water supply pumpers to run dry. In many cases, a route that increases distance, but eliminates potential conflicts and offers improved road conditions, may prove better than a more direct route.

When you assign individual tankers to a particular fill site, remember that some sites have less maneuvering space than others. So, for example, a lighter, smaller tanker probably could move more water in a restricted area than a much larger but less maneuverable tanker.

Staging areas are useful for minimizing congestion at both ends of the shuttle, and it is helpful to locate them within sight of your operation. If they are too far away, tankers getting into position to dump after being called for are slowed, which can interrupt the water supply.

ORGANIZATION

With their multiple loading and unloading stations, high-capacity water shuttles need a smooth-running organization. Tankers do not always fill at the same site or unload at the same station, so they need to be sent where they are needed with a minimum of delay or confusion.

Among the variables to consider in tanker fill site selection are:

- First, the more the distant the fill site is from the dump site, the

larger the tankers need to be. When one fill site is appreciably farther from the dump site than the other, the larger tankers would probably do a better job of hauling water from the more distant location.
- Second, the larger the tanker, the more important it is to achieve a high fill rate. If one fill site can supply a significantly higher fill rate than the other, it usually is best to send the larger tankers there.

You will find that these principles often conflict. When you can get the fastest fill rate at the closest fill site, distributing the tankers becomes a matter of individual judgment, and needs to be adjusted as the operation progresses.

It is important to remain flexible enough to make changes as they are needed. That could mean rerouting tankers as the shuttle develops. One fill site might work better than another, so send additional tankers there, or modify a dump site if bottlenecks develop. Your needed fire flow may change. A good water supply is one that adapts easily to changing conditions.

SUPPLEMENTAL WATER HAULERS

There are times when additional tankers (supplemental water haulers) are needed to support a high-capacity water supply. There are many commercial vehicles used to haul liquids, but many have no means of loading or unloading quickly, and are not adaptable to the techniques and procedures normally used by the fire department.

The tank trucks equipped and used for filling swimming pools in many rural and suburban areas require very little modification for water shuttle use. They generally use a pump to unload into swimming pools and are able to unload at rates in excess of 500 GPM. They normally are filled from fire hydrants, and their fill connections usually are equipped with National Standard Threads on standard fire department fittings.

Transmit mix concrete trucks also can be easily adapted to water shuttle operations. Depending on their size, these trucks can transport from 1,500 to 2,500 gallons, and can dump water in about a minute by turning the drum and using the chute the same way they normally unload concrete. Spillage is minimized by turning the drum backwards while traveling from the fill site to the dump site. What is problematic is filling them. When you discharge more than 500 GPM through the

Figure 12-15. Tank trucks used for filling swimming pools can be used for supplemental supply.

Figure 12-16. Cement trucks can dump as much as 2,000 gallons of water in less than 1 minute by rotating the drum.

open end of a 2½"- or 3"-fire hose, it creates a great deal of reaction. Since the fill crew has to climb a ladder to insert the hose into the drum, they are working from an unstable position, and filling with a normal fill line is very dangerous. One solution is to construct an overhead fill line that discharges directly from the pipe into the drum. This is an ideal way to fill mixers and minimize the fill time. On the other hand, it also limits fill locations, and might preclude mixer use in other parts of the fire management area.

Another approach, and one that permits filling mixers at any adequate water source, is to construct a fill device that can be connected to the end of the fill line. It should include a diffuser on the end, similar to the fill device used for filling fire department tankers from the top, and a means of fastening it to the vehicle before the fill line is charged.

You also can fill a mixer by placing a 10-foot section of 2½"-hard suction hose on the end of a fill line with a diffuser on the discharge end of the suction hose. You then insert the suction hose into the drum, and tie it to the ladder with a rope hose tool or specially designed strap. Using transit mix trucks for water shuttle operations requires that you construct a fill device, and give drivers additional training.

One of the water supply officer's responsibilities should be to survey the fire department's territory for potential water haulers, and then make agreements with their owners to use them in emergencies. Procedures for activating the emergency water haulers, and arrangements for compensating owners and drivers, need to be formalized and standardized before the water haulers are needed.

CONCLUSION

When water supply officers control operations closely, and make solid arrangements for needed apparatus, equipment, and personnel, shuttles can provide the fire flow needed to handle an emergency, wherever and whatever it is.

13. Specifications for Mobile Water Supply Apparatus

OVERVIEW

It is unusual for a fire department to purchase a new pumper with a pump capacity of less than 1,000 GPM; many are rated at 1,250 or 1,500 GPM. A new pumper is purchased after detailed planning and the writing of specifications that stipulate that its pump is adequate to supply the flow required for the area it protects. How to supply water for the new pumper tends to be a secondary concern.

Fire pumps function only to move water from one place to another. Given that a typical pumper's water tank is 500 gallons, simple mathematics show that a 1,500-GPM pumper could operate at its maximum flow for only 20 seconds from a tank of this size. With an empty tank, a fire pump serves no useful purpose until additional water becomes available.

MOBILE WATER SUPPLY APPARATUS

While rural fire departments have taken advantage of Mobile Water Supply (MWS) apparatus for some time, suburban and urban fire departments are becoming more aware of its capabilities and the flexibility it offers to fire department operations.

Most early tankers used by fire departments were converted secondhand commercial vehicles. For example, when a petroleum product distributor replaced a delivery truck, the company donated it to the local fire department to haul water. Milk trucks, liquefied petroleum gas delivery trucks, and military surplus vehicles all have been modified and used for mobile water supply. Many of these units have not been converted safely, and some have been involved in serious accidents. Water weighs considerably more than the product the vehicle was originally designed to haul, and it becomes overloaded when used as a tanker by the fire department. Inadequate brakes, springs that are too light, tanks not securely mounted on the chassis, undersized tires, inadequate or nonexistent baffling, and underpowered engines all contribute to vehicle instability, making some conversion units too dangerous to drive on the highway.

Without a proper conversion, a tanker cannot transport enough water to be useful. For example, using an original fuel oil pump limits the transfer rate to less than 100 GPM in many cases, not enough water to sustain more than one 1½"-line, and certainly not enough to support an effective initial attack. Even with modifications the fill and dump rates might be too slow to provide a dependable water supply.

Although recently many conversion units have been replaced by specially constructed tankers purchased for mobile water supply, many of the same problems typical of conversion units are still evident. A lack of uniform standards in the industry means that some tankers are simply better than others. With few accepted standards, very expensive apparatus can be relatively ineffective, while much cheaper tankers are able to move large amounts of water very efficiently.

In 1991, the National Fire Protection Association (NFPA) adopted 1903, *Standard for Mobile Water Supplies*, which deals solely with tankers and their construction. Unfortunately, much of NFPA 1903 is based on traditional design principles and the capability for fire attack and not on the techniques needed for high-capacity water shuttles. The technology is available to build tankers that are much more efficient than this standard envisions. A tanker built to NFPA 1903 generally would be well designed and constructed to supplement initial water supply for an attack pumper. However, it might not be adequate for a high-capacity water shuttle operation. The only way to get a tanker that is suitable for your fire department is to write detailed specifications for the manufacturer that builds the apparatus.

WATER SHUTTLE OPERATIONS

Many rural and suburban fire departments have increased the tank size on their attack pumpers to get more water for initial attack. Gaining in popularity are pumpers with 1,000-GPM pumps and 1,000-gallon water tanks. While an improvement over the standard 500-gallon tank, 1,000 gallons will still supply the pump for only one minute operating at its full capacity.

Other departments use semi-trailer units to deliver as much as 6,000 gallons on initial response, theoretically enough to handle any single-family dwelling, and attack some larger fires. It would, however, provide only six minutes of water for a 1,000-GPM pumper operating at its maximum capacity. You can see that no fire department could expect to handle a major fire with the amount of water on its first-in apparatus. The capability to shuttle water from a distant water source to a fire scene is the only practical alternative when the closest water supply is too far away for direct hoselines.

APPARATUS SELECTION

Your first step in writing specifications is to decide what you expect the apparatus to do. Pumper tankers, water supply tankers with a fire pump, and general purpose tankers, defined in Chapter 5: Testing Water Supply Apparatus, are multi-purpose units, and while their versatility might look attractive, a piece of fire apparatus designed for several purposes rarely performs any of them as well as a unit designed to fill a specific need.

Pumper Tanker

Many departments have replaced their standard pumpers with pumper tankers because since the tankers operate as mobile water supply units, there is no need to add another vehicle to their fleets. When dispatched, a pumper tanker carries enough water for quick fire attack before an actual tanker arrives, a significant advantage when limited staffing makes it difficult to put a second unit in service. Then, when additional water does arrive, a well-constructed pumper tanker serves both as a reservoir on the fire scene and as a nurse tanker. One pumper tanker can serve as an attack pumper, supply pumper, and nurse tanker, all at the same time, simplifying water supply operation setup.

It may be cost-effective to combine two units into one. Lower insurance premiums, less maintenance, and reduced storage space requirements all are benefits. All things considered, however, the anticipated savings may be more theoretical than actual.

A typical fully equipped pumper with a 500-gallon water tank weighs nearly 30,000 pounds. A larger tank adds approximately 10 pounds for each gallon over 500. A pumper tanker that carries more than 1,000 gallons of water typically needs tandem axles on the rear to handle the additional weight. Many pumper tankers carry 2,000 gallons of water or more, and weigh more than 50,000 pounds. It is costly to provide a chassis heavy enough to handle this much weight, and a drive train powerful enough to perform acceptably. Also, a vehicle heavy enough to carry this load is difficult to drive and probably not maneuverable enough to get where it is needed most. Given the higher maintenance costs for more powerful engines, transmissions, and other drive train components needed to handle the extra weight, it may be cheaper to buy two separate vehicles—a pumper and a tanker—than to combine both functions.

Water Supply Tanker with a Fire Pump

Fire departments often try to gain greater additional flexibility by adding a standard fire pump to a water supply tanker (WST-P). You add a pump primarily so it can set up a draft and load itself and discharge through standard supply lines for direct supply to an attack pumper. Adding a pump gives you other options. The WST-P provides additional pumping capacity so the unit can supply water in various ways and shuttle water. Use it as the source pumper for a relay when direct supply to the scene of an emergency is practical. Use it at a hydrant, when one is available, and you will not have to commit a fully equipped pumper for this purpose. It is problematic to use it as an initial attack unit, because since it was designed as a water hauler, the WST-P typically has a large water tank, but lacks compartment space and hose beds. On one hand, the weight of too many equipment compartments could overload the chassis, cause maintenance problems and make the unit difficult to drive, but, without additional compartments, it may lack the equipment and hoselines for an effective fire attack that would take advantage of the extra water it carries.

Water supply tankers generally carry at least 1,500 gallons of water, and sometimes as much as 4,000 gallons. If you keep compartment space and equipment to a minimum, it is possible to carry as much as 2,500 gallons of water on a two-axle chassis. Most combination units

Specifications for Mobile Water Supply Apparatus

Figure 13-1. Above, Pumper tanker carrying 2,413 gallons of water weighs 60,500 pounds. Below, WST-T tanker carries 4,230 gallons of water and weighs only 58,200 pounds.

Figure 13-2. WST-P is equipped with a 750-GPM pump and carries 1,842 gallons of water, weighs 38,340 pounds.

with a capacity of more than 1,500 gallons have tandem rear axles. Keep in mind that high-capacity WST-P units are also large, heavy, and expensive, like the pumper tankers, and have many of the same advantages and disadvantages.

General Purpose Tanker

A general purpose (GP) tanker may be the best solution to the water supply problems that confront rural fire departments. A general purpose tanker usually is built on a two-axle chassis, and, with careful planning, you can keep its gross weight under 35,000 pounds. With a water tank of 1,200 to 1,800 gallons, the unit can carry a limited amount of the equipment an engine company needs to make an initial attack. The booster type Power Take-Off (PTO) driven pump provides enough water for initial attack, and can supply up to 500 GPM directly to an attack pumper.

General purpose tankers are especially popular with small volunteer fire departments that often can staff only one piece of apparatus at a time. With a GP tanker, two or three firefighters can make a quick, but somewhat limited attack on a fire, either knocking it down or controlling it, before mutual aid units arrive. When it is functioning as a mobile water supply, use the pump to boost pressure to a jet dump to discharge water at a high rate of flow. Alternatively, a large gravity dump can discharge as much as 2,000 GPM. With good planning and proper design, this vehicle can be reasonably efficient in any of its uses.

Water Supply Tanker

A water supply tanker has only one purpose: to transport water between a water supply point and an incident scene. It needs little compartment space since it carries only the equipment it needs to function as a water supply unit. If it has a pump, it is set up for maximum water transfer, not to develop the pressure needed to supply attack lines.

The primary objective in water supply tanker design is efficient water transportation. One of the most important means of meeting this objective is by achieving the shortest possible loading and unloading times. The quickest way to unload a tanker is by using a large gravity dump valve or a jet-assisted dump outlet. Direct fill lines into the tank offer the best loading time, and multiple fill inlets give it the flexibility to function in any type of water shuttle operation; for example, a nurse tanker serving as a reservoir at the dump site, or as a water hauler.

DETERMINING NEEDS

It is on the basis of a detailed needs analysis of the area to be protected and the type of operation envisioned that you decide which type of tanker to build. Combination units are useful for quick attack and are a good choice if:

- You need a large flow for initial attack. With a pumper tanker's large water tank, it can supply the initial flow needed to control a fire quickly;
- You are experiencing a delay and need an interim emergency water supply before your planned water supply is established; or
- You need to supplement an inadequate water system for initial attack.

When the initial response apparatus is used this way, the second alarm assignment (water supply tankers or mutual aid units) must set up a water supply to provide needed flow for a sustained period since the already committed first-in apparatus would be unavailable to shuttle water or set up a relay.

Rural and suburban fire departments might find general purpose tankers useful. The pump should both support an initial attack line and be able to supply a backup 2½"-line. There is sufficient compartment space to carry all of the equipment necessary for a quick attack, and the hose bed can carry enough supply line for initial attack and to make the transition to a sustained water supply.

This type of apparatus also is light enough, small enough, and maneuverable enough to function in tight quarters, travel over narrow roads, and cross bridges with low weight limits safely. If it is used as a "first-out" unit, it must be replaced by a pumper when one arrives, or depend on mutual aid units for additional water as the initial supply is exhausted.

The general purpose tanker is very effective in water shuttles. Properly designed, the combination of quick turnaround time and maneuverability because of its relatively small size and weight give it flexibility under difficult conditions. A high percentage of usable water and quick handling time provide a good water supply.

Water supply tankers give you the most water for the least money. An economically designed (simple) water supply tanker can haul more water than a comparably sized combination unit, and does so with a lighter chassis and drive train. With careful body and tank construction and limiting the compartment size and the amount of equipment it carries, you can put as much as 2,500 gallons on a two-axle chassis, 4,000 gallons on a tandem axle, and 6,000 gallons on a semi-trailer rig. Because of the amount of water it can haul and its lower price, a water supply tanker generally represents the lowest cost per gallon. It may be possible to buy as many as four water supply tankers for the price of one combination pumper tanker.

On the other hand, in minimally staffed departments which have limited space in the fire station, or special fire protection problems, a combination unit can be a good choice. In most situations, a water supply tanker, used in conjunction with well-equipped attack pumpers (those that carry 750 to 1,000 gallons of water), represents the most cost-effective way to sustain an adequate flow when large areas are without hydrants, or to supplement an inadequate system.

DETERMINING TANK SIZE

When you are trying to decide how much water a mobile water supply apparatus should carry, ask: What is the terrain? What types of roads and bridges must be traversed? How large is the service area? How much fire flow potentially is needed? How much staffing do I have to operate it? How much money can the department afford to spend?

Weigh the answers to these and other relevant questions, before you write your specifications. Consider the following:

- First, the farther the tanker must travel in a typical water shuttle, the more water it must carry. On short runs, small, maneuverable tankers may be able to deliver as much water, or more, than larger ones.
- Second, larger tanks mean heavier vehicles. A two-axle pumper tanker typically weighs about 20,000 pounds, plus 10 pounds per gallon of water. A tandem-axle chassis adds another 10,000 pounds. A water supply tanker with a fire pump weighs approximately 15,000 pounds, plus 10 pounds per gallon. A general purpose tanker weighs about the same, depending on the amount of equipment it carries. You can expect a water supply tanker without a standard fire pump to weigh about 10,000 pounds, plus 10 pounds per gallon carried. A tandem-axle, chassis adds approximately 5,000 pounds to the gross vehicle weight of any of these units, except the pumper tanker.
- Third, as weight increases, price escalates rapidly. Two small tankers may be cheaper than one large one.
- Fourth, in rural areas with narrow roads and bridges with low weight limits or mountainous terrain, tanker gross vehicle weight should not exceed 30,000 pounds. Depending on its construction, the tanker will be able to carry from 1,500 to 2,000 gallons of water. If you include the water on the attack pumper, a tanker of this size can supply enough water for initial attacks on single-family dwellings, small barns, and small commercial buildings.
- Finally, you need a water supply tanker with a capacity of 2,500 gallons or more for initial attack, and water shuttle operations for the building sizes found in urban and suburban areas.

The NFPA water supply capability formula is based on maximum tank capacity. During its evaluation, the Insurance Services Office (ISO) reduces the full load capacity by 10% because, according to the law of averages, 10% of the water a tanker carries is unusable in a water shuttle because the tank fills incompletely, or water is left after the tanker is unloaded. Results from hundreds of tanker tests show that the actual amount of usable water varies from a low of 60% to a high of 99.5%. In tanker specifications it is the usable water that is most important, not full load capacity. Baffles constructed inside the tank, location and sizing of vents, and location and arrangement of piping all increase usable water; include them in your specifications.

Vehicle Capacity

Load-carrying capacity is a measure of a mobile water supply unit's capabilities. Especially important is the ability of the chassis and suspension to handle the weight safely. A Gross Vehicle Weight Rating (GVWR) is assigned to each vehicle as it leaves the factory. The GVWR is based on frame strength, the capacity of the springs and other components of the suspension, and on tire size, and braking power. A vehicle's GVWR is limited by the lowest rating of any individual component. Exceed any one rating and the vehicle becomes a safety hazard. If you compare the fully loaded weight to the GVWR, you can determine whether it can handle a load safely. Poor weight distribution often overloads one of the two axles, even when a vehicle's gross weight is within the rated capacity.

Tank Construction

Include in your specifications detailed tank construction information, including construction material, size and shape, size and location of vents, piping arrangement and other details.

Tank Shape

Water tank shape is a prime factor in apparatus highway stability as shown in Figure 13-3. The higher on the vehicle the load is carried, the higher the center of gravity. Generally, flat, square, or "T" type tanks provide the lowest center of gravity, and therefore the best-handling vehicle. A round tank gives the highest center of gravity, and the elliptical tank is somewhere in between. Figure 13-7 shows the effect of the center of gravity on vehicle handling. Whenever a vehicle turns, the load attempts to push it in its current direction of travel, while tire traction on the road's surface forces it to change direction. The combination of the push of the load and the change in direction creates a line of force that can cause vehicle rollover, shown by the arrows in Figure 13-7. If the resulting line of force falls within the vehicle's wheel base and tread width, it stays upright. If it falls outside the wheels, the vehicle may turn over. The higher the center of gravity, the greater the chances the vehicle will roll over.

Round tanks are strongest, and generally suffer fewer stress problems as the tanker ages. Flat tanks give you more space to store hose and equipment, but their surfaces are more prone to leak because of the

Specifications for Mobile Water Supply Apparatus 353

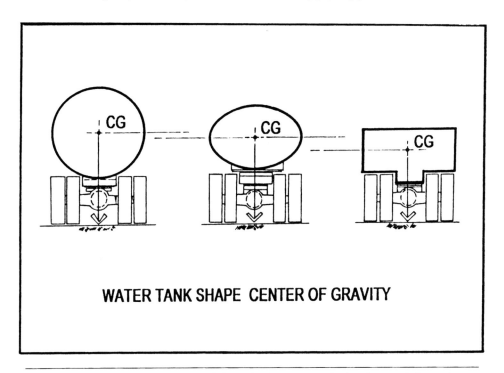

Figure 13-3. The center of gravity of the vehicle varies with the shape of the tank.

Figure 13-4. Round tank on a tanker.

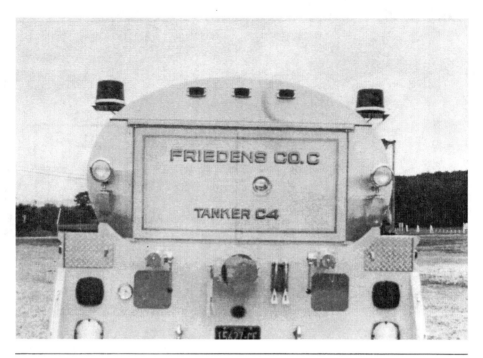

Figure 13-5. Eliptical tank on a tanker.

Figure 13-6. Pumper tanker with a T-shape tank.

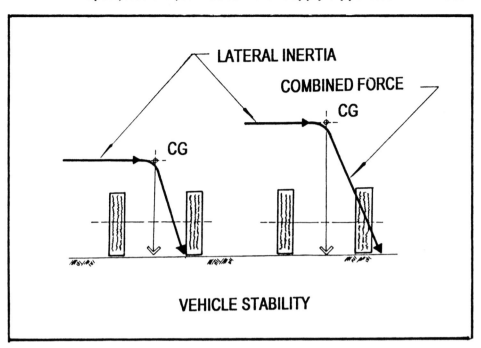

Figure 13-7. The stability of the vehicle depends on the weight, center of gravity, and the tread width.

stress placed on them by vehicle movement, and they are more likely to sustain damage if the tank is pressurized by overly rapid filling.

Vacuum Tankers

A conventional fire department tanker is rated for a 0 PSI pressure differential, and pressurization, or a vacuum, can damage it. A vacuum tanker has a tank that is designed to withstand either pressure or a vacuum.

The tanker's vacuum pump creates a vacuum inside the tank for loading. With an airtight hard suction hose, a vacuum tanker can draft from a static source to fill its tank, and when it dumps, it uses the same pump to pressurize the tank and increase the discharge rate. According to manufacturer information, this tanker loads at a rate of 1,000 GPM from draft and accepts water at fill rates up to 3,000 GPM from a pressurized hydrant. It dumps at a rate of 1,250 GPM, and can develop a limited amount of positive pressure to supply water through large diameter hose directly to an attack pumper over a short distance.

Designed to function only as a water supply tanker, a vacuum tanker does need an additional pump to supply water through 2½″- or

3"-lines to a nurse tanker or directly to an attack pumper or to supply the pressure needed for attack lines.

A vacuum tanker can operate without a water supply pumper to load it, but it would take longer than using a high-capacity fill pumper. A certain amount of time is lost in connecting for draft each time it arrives at the fill site. Probably its best application is in rural settings where the tanker would fill the tank on the attack pumper or a portable tank, then go for another load, where, without needing a pumper at the fill site, it could operate independently. It might be difficult to integrate a vacuum tanker into a high-capacity water shuttle where most of the tankers use conventional gravity or jet dumps to unload.

Construction

The 10-gauge steel tanks traditionally used on fire apparatus had life spans of about 20 years, after which they had to be replaced because of the chemical composition of the water normally carried, and the type of maintenance they previously had. Some manufacturers add cathodic rods to reduce electrolysis and prolong life span, but the rods need periodic replacement and the tanks need maintenance. The weight of a steel tank is another disadvantage. The rule of thumb estimate of 10 pounds per gallon assumes a standard steel tank, but a pressurized tank with stronger construction would increase this figure. Using lightweight materials would decrease it.

Many manufacturers use stainless steel for tank construction because it lasts longer, needs less maintenance, and water stored in it stays clean and uncontaminated. While not a recommended practice, stainless steel tanks can be flushed and used to haul potable water as they were during the aftermath of a hurricane in Charleston, South Carolina, when fire department tankers hauled potable water to areas that were without their normal water supply. A stainless steel tank is somewhat lighter than a standard steel tank.

Lighter still are aluminum tanks. While aluminum does not deteriorate as quickly as steel, it is susceptible to electrolysis and certain types of chemical action, so for fire service use, manufacturers typically coat the inside of the tank to preserve it and lengthen its useful life. Aluminum lacks the strength of steel, and an improperly constructed tank is more susceptible to leaks and structural damage.

Fiberglass tanks have been used for pumpers, and in some instances tankers, for some time. Fiberglass tanks are lighter than aluminum, and the added weight of water is closer to 9.5 pounds per gal-

Figure 13-8. WST-T with aluminum tank carries 2,013 gallons of water and weighs 28,757 pounds.

lon than to ten. Fiberglass tanks have long life spans, and suffer little deterioration. Some manufacturers give lifetime warranties for fiberglass tanks. Fiberglass tanks must be mounted correctly, and supported, to prevent damage and ensure reliable service.

Fire apparatus manufacturers recently have been offering polypropylene nitrogen-welded or molded tanks. Similar in weight to fiberglass tanks, these plastic tanks are lighter than metal tanks, and won't rust or corrode. Most carry a 20-year manufacturer's warranty. Properly constructed and mounted, a polypropylene tank should last for the life of the vehicle.

Some polypropylene tanks carry manufacturers' warnings that exceeding a 1,000-GPM fill rate could damage the tank. Depending on the venting arrangement, this limitation could affect other types of tanks, but not tanks of less than 1,000 gallons, because their fill times of one minute are more than adequate. Even if a tank meets NFPA 1903, *Standard for Mobile Water Supplies*, if its size exceeds 1,500 gallons, the fill rate limitation increases fill time, and could interrupt flow during a water shuttle with high-capacity quick-dump tankers. Inadvertent damage can occur when high-capacity pumpers used to fill tankers from draft, or fire hydrants with capacities greater than 1,000 GPM, boost the flow higher than 1,000 GPM without anyone realizing it. The manufacturer's warning not to exceed a 1,000-GPM fill rate could be problematic if a tank is damaged in service. With no way for operators to determine the actual fill rate during tanker loading, manufacturers could claim that they exceeded the permissible fill rate, and be reluctant to honor the warranty. It is simpler, and more efficient, to specify a tank able to tolerate a fill rate of at least 1,500 GPM, or the peak flow available from a hydrant.

Both fiberglass and polypropylene tanks offer flexibility in their construction, but because they are difficult to change later, original specifications should include the details of venting, baffling, maximum fill rate, and all needed external connections.

Tank Compartmentation

Anti-swash partitions, also known as baffles, are necessary to limit water movement inside a tank. Water is dynamic, and provides resistance when a vehicle changes direction or speed. Water movement makes a tanker difficult to handle, and increases its stopping distance. The amount of force exerted depends on how much water is moving, and the distance it moves before it comes into contact with an obstruction.

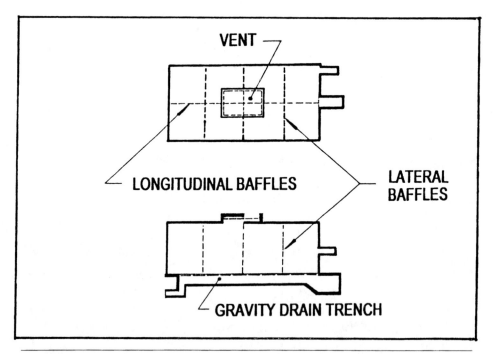

Figure 13-9. Baffles limit the movement of water but the trough under the tank provides for rapid dump and fill. Vent is in the center for most effective air movement with minimum spillage.

Baffles should be spaced close together to limit unrestricted water movement; however, if they are too restrictive, water cannot move between compartments in the tank toward the outlet fast enough to maintain a high dump rate. Too much restriction could cause the pump to go into cavitation, or the flow from the dump outlet to decrease to the point where continued dumping is not worthwhile even though a substantial amount of water remains in some tank compartments. This residual water is known as ballast. Useless, it needlessly increases the vehicle's weight as it moves back and forth in a water shuttle.

You can increase vehicle safety by specifying that baffles be installed to limit unrestricted horizontal water movement inside the tank to no more than 24 inches in any direction. Figure 13-9 shows a typical configuration of baffles that meets this requirement. To ensure free water movement within the tank, you need openings in the baffles at both the top and bottom. The area of these openings in square inches should be at least twice the largest outlet from the tank. Include this as a performance standard, by requiring that at least 95% of the water empties from the tank at its maximum flow rate before either the pump goes into cavitation or the discharge from the dump outlet decreases to a flow of less than 25% of the size of the opening.

In the Figure 13-9 configuration, you can see one way to ensure free water transfer within the tank without disturbing the integrity of the baffles. This tank has a trough that runs longitudinally under the tank. The sump at the front of the trough supplies the tank to pump line, and a large gravity dump is located at the tank's rear. Not only does this trough allow water to flow unrestricted from each compartment to the dump outlet, but the discharge below the bottom of the tank increases the head pressure, yielding a higher dump rate. Figure 13-10 shows the discharge from a tanker built with a trough similar to the diagram. In tests, this type of tanker has yielded dump rates as high as 2,000 GPM.

Venting

Tanker vents serve two primary purposes. First, when a tank is unloaded, air must *enter* the tank to displace the water being discharged; this prevents a vacuum from developing inside the tank. Second, when the tank is filled from an outside source, the air inside the tank has to *escape* as water enters; this prevents pressure from building up inside the tank. With the exception of a few specialized applications, e.g., vacuum tankers, a water tank is not a pressure vessel; it is designed to work with a 0 PSI differential between atmospheric pressure outside and the pressure inside. Either a vacuum or a pressure

Figure 13-10. Dump outlet mounted at the end of the trough below bottom of tank provides flow rates in excess of 3,000 GPM.

buildup inside the tank could damage it. In one case, a fire department accepted delivery of a tanker ordered under specifications that required that 1,500 gallons of water be loaded within two minutes. However, when this department tried to fill the tank at a rate of 1,000 GPM, pressure built up and damaged the tank. The manufacturer replaced the tank at no cost to the fire department, but the problem of meeting the required fill rate without damaging the tank remains.

When 1,500 gallons of water enter a tank, 200 cubic feet of air must escape. You can apply as much as 50 PSI to the fill opening in the tank to force the water in, but your objective is to let the air escape with no appreciable pressure differential across the vent opening. For this you need the largest vents possible. The 1991 editions of NFPA 1901 and 1903 specify that total venting "should be at least four times the cross-sectional area of the sum of all tank fill connections." Your specifications should reference this section of the applicable NFPA standard and you should stipulate that this provision be met.

Vent size is a critical factor in rapid dumping. Tank damage occurs less frequently during dumping than during filling, because any vacuum that develops inside the tank subtracts from the head pressure—the force that causes the water to discharge through a gravity dump.

Specifications for Mobile Water Supply Apparatus

With reduced head pressure, flow decreases and prevents a dangerous vacuum from developing, but increases dump time. If you have a large dump outlet, the vent area should be at least as large as the size of the dump; a vent three times as large would increase apparatus performance. Inadequate venting can decrease discharge rates from large dumps by as much as 50% of the maximum flow.

The vent also can serve as a tank overflow. Most tankers leave a trail of spilled water as they travel to an incident scene. This both reduces the amount of usable water at the emergency scene and, if it spills on the road in front of the vehicle's rear wheels, creates a hazard by reducing the traction of drive wheels. Minimize spillage by locating the vent in the exact center of the tank, the area of least movement, no matter what changes occur in vehicle direction or speed. If there is a hose bed on top of the tank, this may not be a practical location for the vent, but vent shape is not critical. Perhaps the hose bed could be split, and a long narrow vent installed in the center between the two sections.

It is not necessary to have a lid on a vent, but if there is one, there must be a way to close the vent and a way to open it automatically, either electrically, hydraulically, or by air pressure, any time a fill line or dump valve is opened. Off setting baffle plates and covering the

Figure 13-11. Open vent direct into each compartment allows maximum flow.

Figure 13-12. Air cylinder opens vents remotely.

opening with a screen keep debris or foreign objects from falling into the tank without obstructing air movement significantly and without damaging the tank by having it closed at the wrong time. One manufacturer has developed a vent with a float that seals the overflow when the tank is full. A tank with this type of vent needs a second vent, left open, to prevent water hammer, dangerous pressure buildup, or both, when the first vent closes.

Your specifications should stipulate that any discharge from the vents or overflow connection discharge behind the vehicle's rear wheels. Also stipulate that no more than 2% of the tank's total capacity can be lost during road travel for a given distance under specified conditions.

HANDLING TIME

Fill and dump times depend on tank construction and its associated plumbing. Fire departments often specify minimum dump and fill rates in GPM. NFPA 1903 also sets minimum fill and dump rates in GPM. Historically, actual flow rates have not been as important as the time required to complete filling or dumping. A water shuttle depends on a smooth flow of tankers between the fill and dump sites. If one

Figure 13-13. Warning label by tank outlet with manually operated vent.

tanker needs a much longer fill time than others, tankers could be waiting to load at the fill site while the attack pumper runs dry. The same problem could occur at the dump site when the water supply in the portable tank or nurse tanker is exhausted, yet tankers that require excessive time to unload tie up the dump site and loaded tankers wait to get into position to dump. A good objective to set in your tanker specifications is to acquire a tanker that can dump at least 95% of its full capacity in two minutes or less, and can be filled in two minutes or less, where a fill site is capable of supplying enough water to do this.

PUMP INSTALLATION

Most mobile water supply units have pumps. Water supply tankers have been broken down into four categories by pump type:

WST-P   Water supply tanker with a standard fire pump, capable of delivering at least 750 GPM capacity at a net pump pressure of 150 PSI.

WST-A   Water supply tanker with an attack pump, capable of delivering at least 250 GPM at a net pump pressure of 150 PSI.

Figure 13-14. WST-N can dump 1,800 gallons of water in less than 1 minute through 12"-gravity dump.

WST-T   Water supply tanker with a transfer pump, capable of delivering at least 250 GPM at a net pump pressure of 50 PSI.

WST-N   Water supply tanker with no pump, with a large gravity dump for unloading and external connections to allow a pumper to draft directly from the tank.

You can set up an effective water shuttle with a WST-N, but a properly installed pump gives the apparatus greater versatility.

A few tankers, primarily conversion units, use a transfer pump with a separate engine drive. While this arrangement has some advantages, the transfer pump is usually not very efficient. Its maximum capacity is limited by the engine's power, and most are not large enough to unload quickly. Even the smallest tanker must unload at a rate above 500 GPM to meet the two-minute objective; very few separate engine pumps can do this. When you try to use one of these pumps to load the tanker, operating from draft, very few separate engine-driven pumps can deliver more than 200 GPM, and loading time becomes too great for an effective water shuttle.

Any of the other fire pumps—PTO, Front-Mount (FM), or Mid-Ship Transfer (MS)—can be used effectively on a WST-P. For maximum efficiency, use as large a pump as possible; 1,000 GPM is a good standard. Specify the maximum flow rate from the tank to the pump because the size of the tank to pump line or lines can limit the pump's ability to supply its rated capacity from the tank. For many 1,000-GPM fire pumps operating from the tank on the apparatus, the maximum flow is less than 500 GPM because the tank to pump line is too small. Table 13.1 lists maximum flows from various tank to pump lines.

Table 13.1. Maximum Flow with Various Tank to Pump Lines

Size	Type	Typical Flow
2½"	1	250 GPM
3"-Line	1	500 GPM
2½"-Line	2	500 GPM
4"-Line	1	1,000 GPM
3"-Line	2	1,000 GPM

If the tank to pump line is too long, or has numerous 90-degree bends or tee fittings, maximum flow rate is reduced proportionately, often a problem with front-mount pumps. Pump to tank outlet distance, and lack of space to route lines under the cab and through the engine compartment often necessitate the use of numerous 90-degree bends and T-type fittings. Either can reduce maximum flow rate. On the other hand, a very short 3"-line with no bends can increase the maximum flow rate of a typical 400-GPM PTO pump to more than 500 GPM. The easiest way to increase the maximum discharge rate from the pump on a tanker is to reduce friction loss in the pipe between the tank and the pump. When you write pump specifications, include performance criteria that require maintaining a specified flow rate until at least 95% of the tank's maximum capacity has been discharged.

Smaller pipe for the front suction than the intake threads would indicate is used frequently, and the reduced waterway increases friction loss to the point that the front suction might draft less than 70% of its rated capacity. Even operating from a hydrant may produce limited flow from a typical front-suction. Tests of a 1,250-GPM pumper equipped with one of the most popular types of front-suction intakes (a 90-degree bend with a swivel mounted above the front bumper) showed that with a 20-PSI residual pressure on the hydrant, the maximum flow through the front suction was only 850 GPM. Specifications for a pumper or pumper tanker should require that the pump supply 1,000 GPM from draft at a ten-foot lift using the front suction.

Figure 13-15. Reduced waterway on front-suction connection can severely limit maximum flow.

Even if a large gravity dump valve or jet dump could yield a shorter discharge time, there are two practical advantages to having a pump on your tanker. In the early stages of a fire the water tank on the attack engine should be filled immediately after the first tanker arrives, even before you set up the water supply operation. If you use the pump on the tanker for direct supply, you can maintain the flow to the attack lines while you set up the portable tank or nurse tanker and put it into operation. With a WST-N, flow to the attack lines probably would be interrupted while you set up the portable tank, possibly at a critical time in the fire attack.

A pump on a tanker is useful during water shuttle operations. If you are using a nurse tanker for storage at the fire scene, you need a separate pumper to draft from tankers unable to discharge their load under pressure. Even if you are using a portable tank to store water at the dump site, a pump on the tanker is helpful. Tankers that can dump only, either by gravity or with a jet dump, must be within a few feet of the tank, and in a large water shuttle operation the dumping area can become congested. Tankers that can pump off at a rate of at least 500 GPM can be unloaded at a remote station, and deliver their water through supply lines of 100 feet or more. The result? Congestion is minimized, and water supply from the shuttle is increased significantly.

Unless it is cost-prohibitive, a pump on a tanker is a good investment. Some of the newer PTO pumps, designed for maximum water transfer rather than high-pressure operation, can supply as many as two 2½"-lines flowing up to 250 GPM each. This type of pump keeps costs down and does not increase the size or weight of the vehicle significantly.

LARGE DUMP OUTLETS

Large dump outlets on tankers (and portable tanks as reservoirs for them to dump into) make it possible to sustain flows of 1,000 GPM or more by water shuttle. A large dump outlet is one that ranges in size from 4 1/2 to 12 inches. The average flow rate from a large dump outlet depends on tank size and depth, baffles, vents, outlet location, and how it is mounted on the apparatus.

Large gravity dump outlets use the head pressure of the water in the tank to force the water through the opening. The lower the outlet is positioned on the tanker the more head pressure, and the greater the flow rate through the outlet. The bottom of the dump fitting must be higher than the portable tank; this limits your installation options. A typical folding tank is 30 inches high. To allow for variations in terrain, specify a minimum clearance of between 32 and 36 inches above the ground at the bottom of the dump outlet when the tanker is fully loaded, depending on which type of tank will be used. Even for "low-profile" portable tanks, the bottom of the dump should be no lower than 32 inches above the ground. Since you must have mutual aid for effective water shuttle operations, tankers should be able to dump into any standard folding tank.

For any given tanker, the larger the dump size, the faster the tank can be emptied. Table 13.2 gives typical average flow rates for dump valves of various sizes.

Table 13.2. Gravity Dump—Typical Flow

Size	Type	Typical Flow
4½"	Round	450 GPM
5"	Round	550 GPM
6"	Round	650 GPM
8"	Round	1,200 GPM
10"	Round	1,800 GPM
12"	Round	2,300 GPM
10"	Square	2,000 GPM
12"	Square	2,500 GPM

(Some well-designed tankers can exceed the typical flows listed for the dump they are using, but for most tankers you can expect these flow rates.) These estimated flow rates assume baffles adequate to permit water to move freely through the tank to the outlet, and a venting system that does not restrict air movement into the tank. Try to avoid the mistake shown in Figure 13-16, i.e., positioning the dump valve on the end of the tank. This reduces head pressure by approximately .5 PSI, as much as 25% in many cases, and the outlet stops running a "full pipe" when the water level in the tank drops below the top of the dump opening.

Instead, mount the dump valve below the bottom of the tank by using a section of pipe connected to an opening in the bottom of the tank. Keep in mind, however, that the friction loss in the pipe limits the maximum flow rate from the dump, negating the advantage gained from increased head pressure. You can increase flow rate and reduce ballast with the configuration shown in Figure 13-9. If you add the trough to the bottom of the tank, the baffles will not interfere with water transfer to the dump valve through the waterway mounted beneath them. Putting a baffle plate over the opening in the bottom of the sump, another type of anti-swirl device, keeps the pump from drawing a whirlpool and going into cavitation before the tank is empty.

Most dump valves are operated manually with the handle mounted on the valve housing. This can be dangerous because one person at the dump site has to get close enough to the handle to open the valve before dumping, so in most cases someone has to get between the tank and the tanker to reach it. Avoid this at all costs! One way is to include in your specifications a requirement for a remotely operated dump valve control. Different manufacturers offer various options, including air pressure or electricity to control the valve remotely, either at the side of the body or from inside the cab. One manufacturer offers a 10"-square dump valve to mechanically extend the valve handle to the edge of the body where it can be operated safely (Figure 13-19). An alternative to remote control is to add a back step to the tanker; this provides a safe working position for dumping and makes it more convenient to work around the unit. Your specifications should stipulate a safe working location from which to operate the dump valve.

JET DUMP

You can increase the average rate of flow from a dump valve by using a portion of the pump's capacity to speed the flow of water

Figure 13-16. Dump valve mounted high on the end of the tank has 200 gallons of ballast, 10% of the full load.

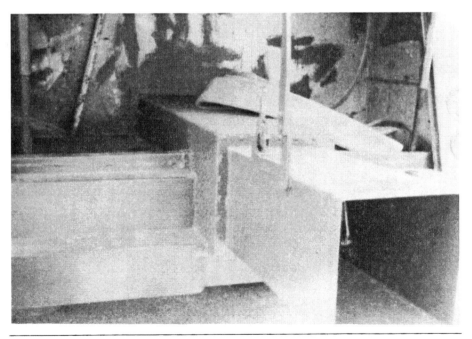

Figure 13-17. Dump valve extended below the bottom of the tank reduces the ballast to less than 1% of the full load.

Figure 13-18. Air cylinder used for remote control of dump valve.

Figure 13-19. Dump valve handle extended to side of the vehicle.

through the dump outlet. Adding a jet assist to dump valves up to 10" in diameter nearly doubles flow rate and cuts dump time significantly. The degree of improvement depends primarily on jet installation and on how the piping to supply it is arranged. While some manufacturers have delivered tankers with jet assists that improved flow rate less than 25% over the gravity dump rate, your specifications should require an increase in flow rate of at least 75% over the average gravity flow rate when the jet is operating.

One advantage of a jet dump versus a large gravity dump is that you can use extension pipes connected to the dump outlet to reduce maneuvering time and your tanker can dump into a portable reservoir from a distance of as much as 8 feet from the tank. Tankers dumping with extension tubes are shown in Figures 13-22 and 13-23. Extensions with 90-degree bends allow you to vary tanker position for greater flexibility in a high-capacity water shuttle. With a properly designed and installed jet dump, extension tubes have little or no effect on the dump rate; overall roundtrip time is reduced by simplified maneuvering and positioning of apparatus at the dump site.

The biggest disadvantage in jet dump use is operational. To use a jet assist, you must put the pump in gear, set it to the proper pressure, and open the valve to the jet assist line. This is difficult for many pump operators and time-consuming. Some manufacturers use such complicated valve arrangements to put the jet into service that operators have thought the jet was operating when it was not. Once the tanker is in position to dump, the gravity dump involves nothing more than operating a handle. There is no need for the driver to leave the vehicle's cab. Simplicity of operation, and eliminating the driver's need to leave the tanker reduce handling time significantly.

FILL LINES

Being able to fill a tanker quickly is just as important as being able to dump it quickly. To do this, you need very large dump valves, because only the tank's head pressure, or the atmospheric pressure plus the head pressure and the velocity of the water when a jet assist is used, is available to move the water. You can use smaller inlet openings for filling, because the pumper can supply up to 150 PSI for loading. If your objective is to load the tanker within two minutes, the larger the tanker, the larger the fill lines need to be. Table 13.3 shows the types of fill inlets needed to meet this objective.

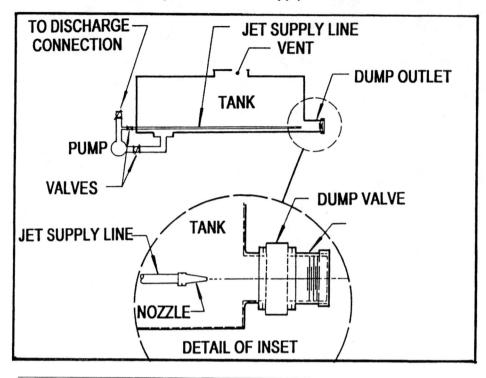

Figure 13-20. Jet nozzle supplied from fire pump increases the flow through the dump outlet.

Figure 13-21. Jet increases the velocity of the water leaving the dump outlet.

Specifications for Mobile Water Supply Apparatus 373

Figure 13-22. Extender tube used with jet dump reduces maneuvering time.

Figure 13-23. Extender tubes with 90-degree angle permit dumping from various positions.

Table 13.3. Tank Fill Connections

Tank Capacity	Fill Connections	
	Number	Type and Size
500–1,500 gallons	1	2½″-NST
1,500–2,500 gallons	2	2½″-NST
More than 2,500 gallons	1	4½″-hydrant threads for soft sleeve
	1	4″- or 5″-quarter-turn fitting for LDH
	3 or 4	2½″-NST

If you close the valve at the tank fill end of the hoseline before the valve at the supply end, the fill line remains pressurized, which could result in injuries when tanker connections are broken. A requirement in your specifications that all fill lines be equipped with a check valve (which closes automatically when pressure is released on the fill line), rather than a manually controlled gate valve, not only provides a greater degree of safety, but saves tanker fill time. You also can save time by installing quick-connect or quarter-turn fittings on all tank inlets.

Direct fill lines at the rear of the tank are useful for water shuttle fill site operations, and smooth the operation when the tank functions as a nurse tanker. With multiple fill lines more than one tanker can discharge its load into the nurse tanker at the same time. They reduce handling time for tankers in the shuttle by allowing subsequent tankers to make their connections and prepare while preceding units are still loading. When you locate fill lines at the rear of the vehicle, extend an additional indicator for the tank level gauge to the rear of the apparatus so that fill site operating personnel can function more effectively.

A combination pumper tanker also should have a tank fill line that can be controlled from the pump panel. If incoming tankers pump directly into the tank, the fire pump is kept isolated from changes in incoming pressure as tankers come and go in the shuttle. This results in a more consistent pressure being maintained to attack lines. If the pump operator observes the tank level gauge closely, and uses the valve on the fill line to keep the tank full, he or she can maintain a consistent discharge pressure without interrupting flow. Use this direct tank fill line the same way when receiving water from a relay as a variation on the open relay. Connect the supply line from the last relay pumper to the direct tank fill line. The tank isolates the pump on the attack

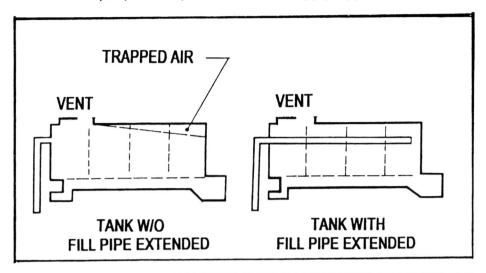

Figure 13-24. Tank fill lines should discharge into the tank at opposite ends from the vent to prevent air pockets.

pumper from excessive pressure from the relay pumpers and pressure surges to the attack lines.

Most tanks are vented on one end or the other, not in the middle where venting is most effective. If the fill line discharges into the end of the tank where the vent is located, air gets trapped inside at the other end, creating an air pocket that causes water to overflow before the tank is filled (Figure 13-24). If you cannot change the vent, your only option is to extend the fill line to the other end of the tank, and drive the air out of the vent as the tank fills. Consider specifying that the tank be filled to at least 99% of its total capacity at the maximum fill rate to avoid this problem.

CHASSIS AND DRIVE TRAIN

Since a tanker's primary function is to haul water, it must be heavy enough to do this safely and efficiently. First consider the Gross Vehicle Weight Rating (GVWR) of the chassis as it was constructed. From a safety standpoint, tanker brakes, tires, and suspension are extremely important. As emergency vehicles, tankers frequently are operated at relatively high speeds, and must be maneuverable. Choose a chassis that has a GVWR of at least 10% above maximum anticipated weight. This margin of safety can make the difference between making a timely arrival at the fire scene, and getting involved in an accident that could cause property damage or life loss.

Vehicle balance also is important because if the front wheels carry too much, or too little, of the weight, the steering will not be responsive and the vehicle might be difficult to control. A bad front-to-rear weight ratio also can cause one axle to be overloaded and the other to be lightly loaded. In extreme cases, the individual axle's weight can be exceeded while the gross vehicle weight remains within the specified limit. In your specifications, require that the front axle carry no more than 40% of the weight, or less than 30% when the vehicle is fully loaded. If it has tandem axles on the rear, the front axle should carry between 20% and 30% of the total weight.

Emergency vehicles generally have the horsepower, and accompanying drive train, to ensure a high level of performance and a speedy response in emergencies. Tankers do not need to respond as quickly as initial response units do because the water an engine company carries should be adequate for a while. At the same time, to be effective, water shuttle operations need tankers that can travel at reasonable speeds, and handle well on the road. For this, tanker engines and drive trains need to be somewhat heavier than similarly sized over-the-road vehicles, but not as powerful as typical pumpers would be. When you specify the chassis, add all the heavy-duty options—larger alternators, extra cooling, heavy duty suspension, etc.—that are available. Many of these options cost little to add and they pay for themselves many times over the life of the apparatus.

ACCEPTANCE TESTS

Do not take delivery of a vehicle without conducting acceptance tests to be sure that the vehicle is constructed in accordance with your specifications. These tests should be conducted in conjunction with the manufacturer's representative, and observed by fire department representatives who know the provisions of the contract. Make sure that the tests are comprehensive enough to demonstrate that all of the specifications have been met. If your fire department lacks the expertise and experience to evaluate the results, hire an outside firm to oversee the testing procedures.

DEVELOPING SPECIFICATIONS

There is nothing standard about fire department mobile water supply apparatus. Each unit is constructed to meet the needs of individual customers and is built according to the specifications the customer provides. Given the cost of apparatus, and the critical nature of the service it provides, it makes sense to develop a detailed set of written specifications for a vehicle's construction and attach it to the contract. Fire department personnel who know little about water supply apparatus or lack the technical expertise to write specifications should hire an outside consultant. It will be money well spent.

14. Standard Operating Procedures

ESTABLISHING STANDARD OPERATING PROCEDURES

Fire department water supply operations generally involve more than one engine company, fire station, or department. For them to work together effectively, they need Standard Operating Procedures (SOPs). Departments that normally respond together need to use the same SOPs.

It takes detailed planning to develop usable SOPs. Each entity needs to know its role and responsibilities. Any officer who might be charged with directing an incident, or any portion of incident operations, must know all of the SOPs, where they will be used, and each unit's role. All officers, pump operators, and firefighters need in-depth training in the use of the SOPs.

Given the substantial role of mutual aid in major fires and emergencies, mutual aid departments have to be included in training efforts that involve the use of SOPs. One of the best ways to promote the use of SOPs is to conduct periodic drills in various areas, using all apparatus and personnel that normally would be dispatched on receipt of an alarm. During the drills, you can identify weak points and refine the procedures. These practical exercises are an opportunity to learn how to maximize available resources. Fire Department Water Supply should always be included in the SOPs. In addition to the incident command structure, a separate water supply organization should be a part of the procedure, and used in every situation where it is applicable.

FIRE MANAGEMENT AREAS

Each fire department area of responsibility should be divided into fire management areas (FMAs). In individual FMAs most of the buildings present similar problems, and similar conditions exist. The base fire flow throughout each FMA should be approximately the same, and the same fire departments respond to each alarm. Individual target hazards within a FMA may require a higher fire flow than the rest of the area; additional apparatus then are assigned to meet special needs.

Alarm assignments should be standard for each FMA and records at the dispatch center should designate the apparatus for each alarm and include information for subsequent and special-purpose alarms.

Another factor in the creation of a Fire Management Area is water supply availability. If a FMA is served by public water systems, it should have complete fire hydrant coverage. In portions of the FMA where static sources are the primary means of water supply, boundaries should be drawn to exclude all areas served by hydrant systems.

The territory of a FMA is limited by the response time of the nearest fire department(s). If alarm assignments change and different fire stations must respond because of distance or accessibility, the FMA designation also will change.

NEEDED FIRE FLOW

All of the methods of determining needed fire flow require preplanning. The Insurance Services Office (ISO) recommends a minimum available fire flow of 500 GPM in areas of predominantly single-family dwellings spaced more than 100 feet apart. The National Fire Protection Association (NFPA) in 1231, *Standard on Water Supplies for Suburban and Rural Fire Fighting*, also recommends a minimum fire flow of 500 GPM for single-family residences.

The 1988 edition of NFPA 1410, *Standard on Initial Fire Attack*, suggests that fire departments arrive on scene with two engine companies; these engine companies should be able to put into service two initial attack lines flowing 200 GPM, and back them up with a third line flowing 200 GPM within 3.5 minutes of arrival. The NFPA standard uses a total flow of 400 GPM for these attack lines, but with the type of nozzles generally used, these attack lines could deliver a flow as high as 500 GPM; thereby meeting the minimum fire flow objectives set by both the ISO and the NFPA. If your SOPs are written to meet the fire flow objectives of NFPA 1410, you must include sufficient apparatus, personnel and equipment to supply at least 500 GPM within 3.5 minutes of arrival.

INITIAL ATTACK—HYDRANT AREA

If fire hydrants are available, your initial engine company's standard practice should be to lay a supply line from the nearest hydrant on its way in, leaving one firefighter at the hydrant to make the connection and open the hydrant when the pumper is ready for water.

As hose crews prepare to enter the building, the pump operator, or another firefighter, should disconnect the supply line from the hose left in the hose bed, connect it to an intake fitting at the pump panel, and signal the firefighter left at the hydrant to charge the line. The firefighter at the hydrant can be contacted via a portable radio. If there is no radio at the hydrant, use the apparatus' air horn to signal that all connections have been made, and the pumper is ready for water. You can open the hydrant faster by putting a hose clamp on the supply line where it comes off the pumper as soon as the vehicle stops moving. Open the hydrant while the connections are being made at the pumper, and release the hose clamp when ready for water. After the line is charged, the firefighter left at the hydrant reports to the incident scene to help with the fire attack.

DRESSING THE HYDRANT

A standard procedure should be to put a gate valve on one or more of the unused outlets of a hydrant before it is opened. Then, as additional engine companies arrive, other lines can be connected. Four-way hydrant valves are useful for the next arriving pumpers to connect into the line and increase the pressure without interrupting the flow to the attack pumper. To expedite this process, consider carrying the four-way hydrant valve, preconnected to the supply line, on the rear step of the attack pumper, as well as a short section of 3" or larger hose in a bundle, to drop off. Then, the pumper that connects to the four-way valve can operate from the hydrant.

A clappered siamese inserted between the hydrant and the supply line can take the place of the four-way valve and enable a supply pumper to boost pressure to the attack pumper through the same hoseline without interrupting the flow. The water supply pumper is supplied through a short section of 3''- or larger hose from the gate valve on the hydrant. You can connect one of the discharge outlets on the pumper to the unused inlet to the siamese, preferably through another short section of 3''-hose. When the discharge pressure from the pumper exceeds the flow pressure from the hydrant on the other inlet of the siamese, the clapper valves operate to supply the hoseline from the pumper, instead of directly from the hydrant. Limiting this operation are the residual pressure at the hydrant and the friction loss in the line used to supply the pumper. If you use short sections of 3''- or larger hose to make these connections, generally 15 to 25 feet, fire flows up to 700 GPM are possible from a typical fire hydrant.

If the SOP is written to include using a siamese in the supply line, carry it preconnected to the supply line in the hose bed. A siamese equipped with a carrying handle makes laying out the hose much easier. You can mount a quick-release fitting on the body or back step of the apparatus to store the siamese, or it can be carried on top of the hose bed, preconnected, with a hydrant wrench fastened to the line ready to lay out.

Consider making up a kit to drop off each time a hydrant is used. Include a hydrant wrench, a gate valve, and an assortment of fittings and adapters, and store them in a case or bag for easy handling by the firefighter who makes the connection.

FLOW LIMITATIONS

Static pressure, system flow capability, hose size, and distance from the hydrant to the attack pumper all limit the maximum fire flow available from a pumper. The SOPs you write to ensure an adequate fire flow will be influenced significantly by flow capabilities and hydrant spacing. Color-coded hydrants make it easier to estimate the available fire flow through direct supply, based on the hydrants that are being used and the area being served.

In Table 14.1 you can see an estimate of the flow you could expect through a single supply line of 1,000 feet. A reasonably good hydrant system should have a residual pressure of at least 40 PSI with the flows this table envisions. When this system is used to capacity, residual pressure could drop to 20 PSI, but should not drop more. The second column is a worst-case scenario; it gives the minimum flow you could expect from this 1,000-foot line.

Table 14.1. Expected Flow with Direct Supply from Hydrant

| Supply Line | Residual Pressure at Hydrant | |
1,000 Ft	Flow at 40 PSI	Flow at 20 PSI
2½″-Hose	125 GPM	75 GPM
3″-Hose	250 GPM	150 GPM
4″-Hose	400 GPM	275 GPM
5″-Hose	800 GPM	550 GPM

Hydrants in residential areas tend to be spaced 1,000 feet apart, but typical supply lines are not more than 500 feet long. In commercial and industrial areas, hydrants should be even closer together. Table 14.2 is based on a 500-foot supply line, which should provide a more realistic estimate of the flow that could be provided by a direct line from the hydrant.

Table 14.2. Expected Flow with Direct Supply from Hydrant

| Supply Line | Residual Pressure at Hydrant | |
500 Ft	Flow at 40 PSI	Flow at 20 PSI
2½″-Hose	200 GPM	140 GPM
3″-Hose	285 GPM	200 GPM
4″-Hose	625 GPM	450 GPM
5″-Hose	1,100 GPM	775 GPM

From these figures, you can see that a single line of 2½″-hose will not supply the flow requirements for two initial attack lines. You can supplement the flow from the hydrant initially with water from the apparatus tank, but without additional water, and very soon, the fire attack would be interrupted. While a single line of 3″-hose could supply enough water for small attack lines, it would take large diameter hose to supply enough water from the initial direct lay from the hydrant to add a backup line.

USE OF A HYDRANT PUMPER

If you insert a pumper into the supply line at the hydrant, you can increase the pressure to as much as 150 PSI; this allows for more friction loss in the line. The maximum expected flow from different supply lines then would increase proportionately.

Table 14.3.

Supply Line 1,000 Ft,	GPM with 150 PSI Pump Pressure
2½"-Hose	250 GPM
3"-Hose	400 GPM
4"-Hose	800 GPM
5"-Hose	1,500 GPM

If you expect both engine companies to arrive on the scene within a few minutes of each other, using the second pumper at the hydrant to boost pressure is a good SOP. Using a "two-piece engine company" this way means that the initial attack lines could be supplied by a single line of 3"-hose or a dual line of 2½". The second-arriving pumper's personnel proceed to the scene and assist in the fire attack using hoselines supplied by the initial attack pumper. Since a typical pumper can supply more attack lines than a single engine company can handle, this is one way to use apparatus and staffing more efficiently. However, by using the second engine company this way, there would be no second pumper on scene as a reserve in case of mechanical problems or other emergencies. Dispatch another engine company for this purpose as soon as you commit the second engine company to the water supply function. If your SOP is to dispatch three engine companies on all structure fires, the third company can be available for water supply while the other two handle fire attack.

Split Lay

If hydrant spacing or traffic patterns make it difficult for an initial engine company to lay a complete supply line from a hydrant to an incident scene, a split lay may be the best procedure to use. With a split lay, the initial engine company drops a line at the intersection where it enters a side street or long driveway. The second engine picks up the end of the supply line and completes the line to a hydrant using a reverse lay. In this case, no hydrant valves or other special fittings are needed because the hydrant pumper connects directly to the large outlet, and the full capacity of the hydrant is available to supply it. The personnel on the second engine are dropped off when the supply line is picked up, report to the incident commander, and assist with the attack. After all connections are made, the firefighter who stayed behind to help the pump operator connect to the hydrant proceeds to the incident scene to assist the attack crews.

Using LDH, with its sexless couplings, simplifies a split lay. If you use fire hose with threaded couplings for supply lines, a good SOP to consider is to carry the hose load with a double female or double male fitting connected to the end of the line. Then, instead of needing an adapter to make the connection when the couplings on the two lines do not match, you need only remove the double male or female fitting from one of them. Any time you can save making this connection is very important, because when you use a split lay the attack pumper receives no water from the hydrant until all connections are completed and both pumpers are in place. The incident commander is dependent on the attack pumper's tank water for the initial supply. When this happens, he or she might have to hold crews back until the water supply from the hydrant has been established.

HEAVY ATTACK LINES

Within most FMAs are buildings that need more fire flow than you would use for an initial attack. Barns, small commercial buildings, large multi-story houses, or heavily involved single-family residences often need more than 500 GPM to bring them under control. First-arriving engine companies should be able to apply as much as 1,000 GPM to a fire within minutes of arriving. To maneuver attack lines into a position to put water where it is needed, and to do so with a tolerable margin of safety, requires the following minimum numbers of personnel:

Type of Line	Minimum Personnel
1½-inch handline	2
1¾-inch handline	2
2-inch handline	3
2½-inch handline	3

Based on these estimates, two engine companies could not apply 1,000 GPM to a fire. Instead you would need four 2½″-handlines and 12 firefighters. On the other hand, a single master stream device could supply as much as 1,000 GPM and need only one firefighter to direct a stream that is positioned and ready to flow water. Personnel requirements can be reduced further, and the master stream put into service faster, if your standard procedure is to carry it on the pumper mounted, preconnected and ready to use.

Many rural fire departments do not carry master stream devices because their officers do not believe they encounter enough large fires to make them worthwhile. As a result, these departments have lost barns and other large structures to uncontrolled fire. Wherever there are large buildings with high fire loads, it is essential to have a preconnected master stream device. All pumpers should be equipped with at least one master stream device, and a standard operating practice should be to use it when fire volume outpaces handlines and you have enough water to supply it for a quick knockdown. In rural and suburban areas, where staffing is limited at certain times, this is especially important.

HYDRANT SUPPLY FOR MASTER STREAMS

Your initial attack SOPs should be written to boost water supply immediately after arrival so that you can use master streams. To get a 1,000-GPM or greater flow from a fire hydrant, you need either large diameter hose or multiple lines. With multiple 2½"- or 3"-lines, you'll also need a pumper at the hydrant. Stipulate that pumpers that normally operate in high-risk areas either carry LDH, or be configured to lay dual lines of 3"-hose. Alternatively your SOP could direct the first engine company to lay a single line directly from the hydrant on its way in. The next pumper makes a reverse lay from the incident scene (after dropping off its personnel to assist the attack pumper) as it proceeds to the hydrant to serve as the water supply pumper.

Initial Attack—Tanker Supply

Initial attacks in areas without hydrants may have to depend on water transported by tankers. The more water brought by first-arriving apparatus, the longer the attack pumper can supply water. The SOP must call for a water supply tanker on the initial response to respond to the scene with the pumper, and connect directly to it to provide water until the water shuttle can be set up. This should be a water supply tanker equipped with a pump that can transfer at least 500 GPM; without a 500-GPM pump, you'll need a different SOP.

If your first-due tanker has a large gravity dump versus a pump, the

SOP may call for a portable tank to be set up so the attack pumper can draft and maintain the attack. In this situation your attack pumper must have a gated suction inlet fitting, which allows suction hose to be connected while attack lines are supplied from the tank on the apparatus.

You can write a SOP for portable tank use two ways. Some fire departments carry a portable tank on each pumper, so the engine company can set up the tank and connect the suction hose while waiting for the tanker to arrive. Others carry the portable tank on the tanker; the crew that arrives with the tanker sets it up and connects the suction hose. How you write your SOP depends on the anticipated personnel strength of the engine companies and the tankers that normally respond together, and how much water the attack pumper carries.

After the portable tank is set up and filled, the transition from the tank on the pumper should be only a slight interruption, or no interruption at all, to the fire attack. The larger the attack pumper tank is, the more time the tanker crew has to set up the tank and prepare for the attack pumper to begin to draft. The ISO guidelines for Fire Department Water Supply stipulate that the setup must be complete, and the attack pumper operating from draft, within five minutes of the engine company's arrival. Based on NFPA 1410 (*Standard on Initial Fire Attack, 1988*) standards, an attack pumper with a standard 500-gallon water tank would deplete its water four minutes after arrival. A tank size of 1,000 gallons provides only an additional minute's flow, and the SOP must be written to provide an additional water supply and expedite the transition.

Keep in mind that using an attack pumper to draft from the portable tank assumes that there will be room both to set up the tank, and for tankers to maneuver, at the point of fire attack. If you lack space, you will need alternative procedures.

Modified Split Lay

Tankers can use the same procedure initial engine companies use to make a split lay to a hydrant. The first-arriving engine company drops a line at the road where it will be readily accessible for tankers to unload. Parking lots or other large open spaces are good locations where a water supply engine company could pick up the line, make a reverse lay and set up a dump site.

DIRECT SUPPLY FROM TANKERS

Tankers equipped with 500-GPM or larger pumps can pick up the supply line where the engine company dropped it and pump their load directly to the attack pumper. In a water shuttle for sustained attack, tankers can pump directly to the attack pumper through the initial supply line. In this case, the SOP should specify siamese or gated wye fittings on the end of the supply line; this allows multiple tankers to connect at the same time. What limits using supply line for direct tanker supply are tanker pump size and hoseline capacity from the supply point to the incident scene.

Nurse Tanker Operations

There are two ways to write a SOP for using tankers to store water at the dump site and serve as reservoirs for shuttle tankers: use the pump on the nurse tanker to supply water to the attack pumper, or use a separate water supply pumper to draft from it.

If your initial tanker's pump can supply 500 GPM or more to the supply line, let it stay in place; shuttle tankers can pump and store their loads into its tank while they go for more water. Without a pump that allows the tanker to function as a water supply pumper and an on-site reservoir, you will need a separate water supply pumper.

WATER SUPPLY PUMPER

Your SOP may be to use the second-due pumper for water supply. In a water shuttle situation, this engine company will probably be operating the dump site and be unavailable to help the initial engine company put attack lines into service. When a nurse tanker is used to store water at a dump site, the water supply pumper must connect to a large outlet from the tanker with hard suction hose. If the procedure is to have a pumper draft from the nurse tanker, position the tanker so that the pumper can get close enough to connect to it with a single section of hard suction hose. Even better: set up a portable tank (or tanks) at the dump site to release all tankers to shuttle water.

Portable Tank Operation

When you use portable tanks at a dump site for storage, where you locate the tanks and the water supply pumper is critical. What you need is a SOP that lets you set up a standard water shuttle as quickly as possible with provisions to expand to a high-capacity water shuttle if necessary. Two conditions smooth the transition to a high-capacity supply. First, the supply lines from the dump site to the attack pumper must be large enough to handle the increased flow, or be supplemented with additional lines. Second, the dump site must be large enough to set up multiple tanks, maneuver tankers to fill them, and have multiple dumping stations so that more than one tanker can dump simultaneously, based on each tanker's unloading requirements.

Organizing the water supply, dump site, and fill site as soon as personnel are available to do so, and determining if a water shuttle is required, should be part of the standard operating procedures.

ORGANIZATION

The fire service's Incident Command System rarely defines water supply organization; in fact, it frequently is ignored until the water problem becomes dire. The Water Supply Officer (WSO) is the key to ensuring an adequate water supply. Along with the WSO, you need an organization that can supply the attack pumper quickly and efficiently.

It is the WSO who decides how to best get water, and how to organize the water supply operation. Using standard operating procedures greatly simplifies his or her job. Generally, groups responsible for specific portions of the operation need both their own command structure, and to communicate with each other and the water supply officer. For example, each relay in a multiple-relay operation, with different hoselines and apparatus, should have its own relay control officer. In water shuttles, each dump site and fill site should have its own organization as detailed in Chapter 9: Water Shuttle Operations. With multiple water shuttles, each should have a water shuttle officer in charge and clearly identified; all are directed by the water supply officer.

STANDARD RESPONSE TO ALARMS

Each fire management area should have a standard level of response for each alarm. The area being served, type of alarm, and each company's resources determine the number of engine companies, water supply units, and service/ladder companies due on each alarm.

PERSONNEL

To staff the minimum number of attack lines specified in NFPA 1410, the first alarm assignment needs at least seven firefighters, plus two pump operators, and at least one officer. Given these figures, a minimum initial response should be two engine companies, with at least ten people, for structure fires in any area. Engine companies in many paid departments have fewer than five members per shift. When average staffing levels for engine companies consist of fewer than five members per engine, include an additional engine company. These numbers do not allow for personnel to perform search and rescue, ventilation, salvage, or any of the other jobs that must be done on arrival at the fire scene. If ladder companies or service companies are available, dispatch them on all structure alarms. Their personnel can do these jobs, freeing engine companies to concentrate on fire attack and water supply.

Volunteer response, on the other hand, can range from a high of 40 to 50 people responding for a building fire, to a low of three, especially at certain times of the day. Volunteer fire departments have to plan for adequate response levels based on average levels of response. A volunteer department with two or three pumpers, and personnel to operate them, can send both initial engine companies from one department, and in some cases, from the same station. Limited apparatus and personnel may require that more than one volunteer department be dispatched on structure fires within a fire management area. Mutual aid required to provide an adequate initial response level should be dispatched automatically with the first-due company.

If ladder or service companies are not readily available, additional engine companies can assume their responsibilities. The engine companies then must carry the appropriate equipment in addition to the required engine company equipment.

WATER SUPPLY REQUIREMENTS

In areas with strong water systems and hydrants spaced less than 600 feet apart, it is possible to write SOPs that use LDH to provide the water supply without a pumper at the hydrant. In other cases, the second engine company may have to provide the water supply for the initial attack. Generally, it is more efficient to include an additional engine company on the initial alarm to handle water supply. A third engine company is essential when a water shuttle is the primary source of water, or the water supply involves setting up a pumper to draft from a static source.

If an incident scene is within 2,000 feet of a water source, use a relay. If the distance is greater, a water shuttle is probably more effective. Either way, the initial response must have adequate equipment and apparatus.

STANDARD RESPONSE FOR RELAY

In relay operations, SOPs usually are based on pumpers in the hoseline spaced at 1,000-foot intervals. When a FMA's base fire flow is only 500 GPM, all you need is a dual line of 2½''- or 3''-hose to transport it. But to handle target hazards in the FMA or fully involved structures, the hose should be able to transport at least 1,000 GPM. Table 4.4 shows how to meet these criteria by giving alternative hose layouts for each section of the relay.

Table 14.4. Required Hoselines for a 1,000-GPM Relay

Four lines of 2½''-line
Two lines of 3''- and a single 2½''-line
A single line of 4''- supplemented by a 2½''-line
A single line of 5''-hose

Also, each pumper must have a rated capacity of 1,000 GPM or more. Pumpers dispatched for water supply without sufficient hose need hose from the attack units as a supplement. Do this by dropping a line as the engine company approaches the scene, or pull additional hose from the hose bed by hand while the attack is underway. Or, dispatch extra pumpers if hoselines are too small; the pumpers can be located closer together, reducing the length of the line and the total friction loss between relay pumpers. The drawback? This complicates the relay operation and increases the probability of mechanical failure or operator error.

It is possible to set up the initial response for 500 GPM and depend on a second alarm to reach the 1,000-GPM requirement, but it means a significant delay in applying a master stream to the fire, with slowed fire control and the possibility of increased damage to building(s) and contents.

STANDARD RESPONSE FOR WATER SHUTTLE

To set a response level for tanker operations, you need to know the capabilities of individual mobile water supply units. The objective of departments that operate in nonhydrant areas is to purchase and equip mobile water supply apparatus that helps them provide a good, dependable water supply anywhere in their territory. In setting standard response levels, fire departments have to plan the best way to use available apparatus. The test results detailed in Chapter 5: Testing Water Supply Apparatus, form the basis for this planning effort.

After tankers are tested, each is assigned a GPM rating over a variety of distances. Based on the available water supply points in your FMA, you dispatch enough tankers on each alarm to ensure a continuous 500-GPM flow capability. Along with the tankers to haul it, you include an engine company to set up a fill site and fill them, and you dispatch the engine company on the initial alarm. It sets up the fill site for the tankers then, rather than being summoned after the engine company arrives on the scene and decides that a water shuttle is needed. First tankers often arrive at the water supply location for filling a few minutes after the initial response units arrive at the incident scene. Delays in setting up the fill site result in corresponding delays in tankers returning to the scene with more water. These delays can cause initial attack crews to run dry at the most critical point in the fire attack, perhaps leaving them unprotected inside a fire building.

When tankers are not available to maintain a 500-GPM flow, it sometimes is more practical to limit the desired flow to 250 GPM. This flow will not support an initial attack level that meets established guidelines, but it permits the initial engine company to put two 1½"-lines or one 2½"-line on the fire. Two hundred and fifty gallons per minute, applied within five minutes of arrival and sustained for a period of two hours, qualifies a fire department for an ISO rating under "Fire Department Water Supply." A standard response that lets you meet the basic 250-GPM fire flow with only initial response apparatus and equipment, is better than the water supply capabilities of many fire departments. Regardless, they need a long-range plan to meet the minimum 500-GPM fire flow as soon as possible

SUBSEQUENT ALARMS

In addition to standard response levels for all alarms within a FMA, the dispatch center also needs a list of standard alarms for additional water supply. If you have enough mobile water supply apparatus available, set these alarms up in multiples of 500 GPM. When an additional alarm for water supply is called for, the necessary tankers, as well as an engine company for filling them, should be listed on the running assignment. This list means that the water supply officer does not have to decide which mutual aid tankers to request, and it expedites the dispatch of additional water supply units. In a typical situation, this might mean dispatching three tankers, with a pumper to fill them, for each 500 GPM of additional water supply that is needed.

SPECIAL ALARMS

Each target hazard or special risk within a FMA should be listed on a special alarm assignment record, with the initial assignment calculated to meet the specific risk needs. You base this response on the risk analysis procedures detailed in Chapter 3: Risk Analysis in Fire Protection Planning, of this book. Special alarm assignments give you the flexibility to deal with life safety concerns, hazardous materials, or high fuel loads by dispatching enough apparatus, equipment, and personnel to bring the situation under control quickly. A common mistake is to set up an initial response for major fire hazards that is inadequate.

Later-arriving apparatus and personnel are confronted with a more advanced fire, and could find themselves playing "catch-up" and chasing the fire until the structure is lost. This is not a good use of available resources.

PREPLANNING AS A BASIS FOR STANDARD RESPONSE AND OPERATING PROCEDURES

Standard operating procedures and standard response levels for emergency alarms work only with good preplanning, not of just specific structures or special risks, but of entire areas. Preplans need to be flexible, yet based on quantifiable and verifiable data. Then, firefighters and dispatch can be confident that standard responses will be sufficient to handle anticipated emergencies without unduly burdening mutual aid departments.

15. Water Supply Officer

WHILE MOST PROGRESSIVE fire departments make a practice of designating a Water Supply Officer (WSO) as part of their incident command structure, very few include a WSO in the department's organizational chart. It takes planning, coordination with other agencies, and specialized training to ensure an adequate water supply. To make all this happen, the WSO has two different roles: one is to take command of the water supply operations during an incident and function as a staff officer; the other is to act as a support officer in an administrative capacity on a *continuing* basis. One person could serve in both capacities; alternatively, two individuals could each assume one of the roles.

ADMINISTRATIVE FUNCTIONS

The WSO should perform all water supply prefire planning based on the fire department's detailed and specific plans to provide the needed base flow throughout each of its fire management areas (FMA), and the higher flow rates needed for individual target hazards and special risks.

RISK ANALYSIS

Water supply planning begins when you determine needed fire flow. (See Chapter 3: Risk Analysis in Fire Protection Planning, and Chapter 2: Water Supply Planning.) The WSO's job is to estimate the needed fire flow, and prepare preplans and standard operating procedures (SOPs) to supply it. It is useful for the WSO to have junior line officers assist in preplan preparation; it involves them in the planning process and makes them more aware of the need for an adequate water supply. Preparing a preplan while they inspect a building gives officers and firefighters an opportunity to learn about structures they may have to enter later under adverse conditions.

EVALUATION AND INVENTORY OF SOURCES

Two of the primary purposes of this book are to suggest some guidelines to use to estimate needed water supply, and to offer some ways to evaluate existing water sources. As a water supply officer, use the procedures recommended here to test the fire flow capabilities of the public water system, and dry hydrants and static sources, or other supply sources you may use. After you evaluate them, adopt a schedule for regular testing and maintenance and make sure all work is performed as scheduled. This is especially important if dry hydrants or static sources are involved. The maintenance intervals and records recommended in Chapter 7: Suction Supply as a Water Source, give you a good starting point, but modify them for your conditions. You may find that tests need to be conducted more often than the recommended intervals suggested here, or that you need additional maintenance, so include these changes in your routine maintenance program for water supply points.

LIAISON WITH OTHER AGENCIES

Your fire department, like many others, may not have sole authority for controlling the planning process, testing water systems, or maintaining fire hydrants. If not, your WSO should have a good working relationship with any agencies that do.

PUBLIC WATER SYSTEMS

The WSO's most important relationship is with the agency responsible for the public water system. This organization's primary objectives are to furnish potable water to its users, and generate enough revenue to pay the operating expenses. Fire hydrants tend to get low priority because they do not generate revenues and their use is on an emergency basis only. As the department's water advocate, the WSO maintains relationships with individuals or agencies that can give firefighting water supply the attention it deserves.

The WSO's involvement in the operation of the water system should be in the areas of planning, system operational status, and problem resolution.

First, water supply officers need to involve themselves in planning new systems and extending or modifying existing ones. Otherwise, fire hydrants could be improperly spaced or left out of the system altogether. The WSO helps prepare a comprehensive water and sewer plan for the area and makes sure it addresses fire protection needs. The plan should set minimum distances between hydrants, as hose can be laid, so that needed fire flow can be supplied, and the Insurance Services Office's (ISO) requirements met.

The WSO needs to participate in site plan review for new construction or modifications. Engineers tend to put fire hydrants in the most convenient locations, that is, where they can be installed at the lowest cost, often paying little attention to whether or not they are accessible to fire apparatus. The WSO should try to review all site plans, and make recommendations about hydrant locations.

By staying in touch with water system personnel, the WSO can check the system's operational status and keep his or her department aware of system problems, for example, inoperative fire hydrants, in time to change preplans or SOPs.

Part of the WSO's role is problem resolution. When there are system problems, it is the WSO who needs to know which individuals to call and how they can be reached during off-hours. Ideally, the WSO has enough credibility with the water system's supervisory personnel to ensure that they resolve problems promptly, and inform the fire department of the status of repairs.

PLANNING AND ZONING DEPARTMENTS

Fire departments need to have a substantial role in the planning and zoning process. Often, new subdivisions are approved with little or no consideration given to a fire department's need for adequate water. In terms of long-range planning, the WSO can relate the available fire flow for the projected population density and construction for the entire FMA. The size of water mains, amount of storage, and hydrant placement for additions or changes to the system in connection with new construction should be included in a comprehensive water and sewer plan adopted by the authority that has jurisdiction over the area. Each new subdivision or development should be required to meet the standards of the plan. Where there is no public water system, alternative water sources must be identified and developed to provide needed fire flow. The WSO's input and oversight are essential.

BUILDING INSPECTORS

Even in well-planned developments, it is up to building inspectors to make sure that construction follows approved plans, and meets all provisions of applicable life safety and fire prevention codes. Here the WSO's role is to provide the necessary training to make building inspectors aware of the fire department's needs, including explaining the importance of installing and maintaining sprinkler systems and other fire safety equipment in the recommended manner. With a close relationship, and frequent dialogue, problems between building inspectors and the WSO can be resolved in an atmosphere of mutual trust and respect.

OTHER GOVERNMENTAL AGENCIES

It takes cooperation from a number of government agencies to provide a dependable fire protection water supply. At water shuttles, police personnel can expedite operations and increase safety by providing traffic control at critical intersections, or, in some cases, setting up detours and closing streets to protect firefighters and bystanders. When hoselines must be laid across major thoroughfares, police personnel protect the hose and reroute traffic. In other cases, arrangements must be made to stop trains when railroad tracks are blocked by fire apparatus and hoselines.

Highway department assistance ranges from cleaning up a highway after an accident, to providing snow plows to give fire departments access during bad weather. Of the many ways the highway department can help a water supply officer, probably the most important is keeping the fire department informed of problems or detours. This information allows the WSO to modify preplans and standard operating procedures to include alternate routes.

The U.S. Department of Agriculture's (USDA) Soil Conservation Service (SCS) can help in water source development. Its employees frequently know of ponds, some not readily apparent, that can be developed into good water supply points. Or, when a fire department tries for Fire Department Water Supply credit under the ISO, the SCS might have information on the reliability of potential water sources and provide the certification of dependability the ISO requires for water supply points.

CONTRACTORS AND DEVELOPERS

Although fire department relationships with contractors and developers often are adversarial, they do not have to be. For builders, time is money. Builders find it problematic when fire departments want to make changes that would delay construction. Avoid this by adopting comprehensive plans, including requirements for fire department water supply, and making them available to engineering firms, architects, and contractors, so changes can be agreed to while a project is in its infancy. Often, changes to construction plans, early in the process, improve fire protection without raising construction costs. In other cases, better fire protection results in lower insurance premiums that more than pay for higher construction costs. By working together, the water supply officer and construction industry representatives can save money and improve fire protection.

PRIVATE WATER SYSTEMS

Private water systems, with provisions for fire protection, typically are installed when high-value buildings and facilities with high fire risk potential are constructed in areas where no public water system is available. Some of the private companies in these areas operate their own industrial fire brigades. Brigade members are employees trained to provide fire protection. However, since most of them are equipped only to handle very small fires and minor emergencies, they depend on the local fire department's help for major incidents. It is the water supply officer's job to know how these private systems operate and what fire protection equipment is available, and to make sure that both the system and equipment function properly when needed.

Since sprinklers and standpipes provide the first line of attack, the water system must be designed to provide the needed fire flow for them. Private sector companies generally lack the personnel and expertise to operate a central water system, and usually depend on outside contractors for maintenance. This maintenance typically consists of replacing defective equipment and repairing leaks, but not testing or maintaining the water system. It may be that the fire department's water supply officer is the only person aware of the system's condition and in a position to suggest repairs or improvements. He or she should keep accurate records of this information, as well as the available fire flow, and incorporate any changes into prefire plans.

WATER SUPPLY MASTER PLAN

Part of the WSO's job is to develop a water supply master plan for the fire department's area of responsibility. This includes writing separate plans for each fire management area that show how enough water will be provided to sustain both the base flow throughout the service area, and higher flows needed for target hazards or special risks.

NEEDED FIRE FLOW

Base the needed fire flow on the results of the inspection and risk analysis surveys carried out using the procedures specified in Chapter 3. During its Public Protection Classification (PPC) survey, the ISO will base the needed fire flow on the fifth largest building in the service area. Since this figure is probably not representative of the types of structures in the FMA, your true base fire flow may be significantly less than the needed fire flow set by the ISO. Generally, a 500-GPM base fire flow is sufficient in rural and suburban areas, with additional flow required for large buildings and target hazards.

WATER SUPPLY POINTS

Prepare, and keep current, an inventory of all water supply points. Include all fire hydrants (pressurized, wet, and dry, and storage tanks), as well as all suction supply points usable for drafting. Sources outside the department's service area, but within range of hoselines or tanker shuttles to provide the needed flow within their coverage area, can be part of the inventory. For each source, prepare a Water Supply Point Record with:

- Location and estimated available flow (list expected seasonal or other variations in reliability or accessibility);
- A signed agreement that gives the fire department permission to use it;
- A record of keys or other means of access;
- Details of improvements that have been made to the water supply point to provide access for a fire department pumper;
- A list of special needs such as extra hose, adapters or fittings, etc., that will be required to use the supply point; and
- A record of periodic inspections and tests.

Mark the location of each point on the map in the fire station, and carry a copy of the map on all apparatus. Give copies to mutual aid departments.

WATER SUPPLY APPARATUS AND EQUIPMENT

It is the water supply officer's responsibility to make sure that approved, required supply apparatus and equipment are available and operational at all times. This means that fire pumps on the apparatus have the capacity to supply required maximum flow, and meet the ISO's capacity requirements. The ISO also specifies what equipment should be carried by engine companies and ladder companies, including nozzles and appliances, hose for attack and supply lines, and adapters and specialized equipment.

In addition to the ISO's recommended equipment, the water supply equipment needed for effective water shuttles or relay operations is listed in this book. The WSO should evaluate each suggested item, determine which items are really needed, and present a list of needed equipment to the fire chief or purchasing department, along with the justification for purchase, to include in the department's budget.

WATER SUPPLY APPARATUS SPECIFICATIONS

Specifications for water supply apparatus were outlined in Chapter 13. After evaluating the fire department's needs, the WSO should help plan the purchase of new apparatus to see that water supply needs are met. This is especially important when mobile water supply apparatus must be purchased. It is the water supply needs of an area that dictate the size and construction of fire department tankers. Decide what to buy based on evaluations of needed fire flow, transport distance, and available water supply points.

TESTING APPARATUS AND FIRE HOSE

Acceptance tests of fire apparatus and fire hose, made at the time of delivery, are the way to avoid purchasing pumpers or tankers that do not meet department needs. Merely including an acceptance test in your specifications increases your chances of getting what you ordered. Actually making the test verifies that the apparatus meets your requirements. If you identify shortcomings before you accept the apparatus, changes can still be made.

Annual service tests of all pumps are the best indication that a pump can still do the job it was purchased to do, and is unlikely to fail during prolonged water supply operations. Tests also help you to monitor the condition of pumps, and schedule major repairs in advance. (Chapter 5 covers testing water supply apparatus in detail.)

Annual tests of fire hose are equally important. (Hose tests are covered in Chapter 4.) The WSO schedules hose tests, oversees them, keeps test records, and determines test pressures. When water supply plans specify operating pressures higher than 200 PSI, fire hose should be tested at 300 PSI instead of the 250 PSI specified by the NFPA. This gives operating pressures of up to 250 PSI with a minimum 50-PSI safety factor.

Although the WSO does not actually conduct these tests, he or she should verify that they are done and review the permanent pump and hose test records on at least a quarterly basis to verify scheduling and documentation.

TRANSPORT DISTANCE

Flow capabilities are calculated by determining how far the fire department can transport enough water to provide the needed fire flow. The anticipated transport distance depends on how the water is moved. When you determine water supply coverage, use the base fire flow for the fire management area as a minimum flow, and evaluate the fire hose carried on the first-alarm units, the Continuous Flow Capabilities (CFC) of the tankers, or both, to determine the distance from the source the required GPM can be supplied.

FIRE HOSE FOR TRANSPORTING WATER

Table 15.1 gives an estimate of the maximum flows you can expect from 2½″- or 3″-hose, with a pumper in line every 1,000 feet. The number in parentheses behind each GPM figure is how many pumpers would be needed to deliver the estimated flow to the attack pumper. To deliver 500 GPM at a distance of 1,000 feet from the water source would require a dual line of 2½″-hose. A single line of 3″-hose could deliver 500 GPM over a distance of 500 feet, but it would take dual lines to move the water 1,000 feet. With only 2½″- or 3″-hose available, the transport distance by first-alarm engine companies through fire hose generally would be no more than 1,000 feet. Table 15.1 shows that it is possible to set up a relay that can move water farther than 1,000 feet, but longer relays tend to require more apparatus and fire hose than are available on initial alarm assignments.

Table 15.1. Estimated Maximum Flow in GPM Using 2½"- or 3"-Fire Hose

Distance to Source in Miles (One Way)	Maximum Flow from a Single Line of Hose to an Attack Pumper Without a Relay		Number of Parallel Lines Required to Deliver Flow to Attack Pumper with Relay Pumpers Every 1,000 Feet		
	Hydrant only (50 PSI)	Supply Pumper (150 PSI)	1 line	2 lines	3 lines
2½"-Hose					
500 feet	225	390	390(1)	780(1)	1,170(1)
1,000 feet	160	275	275(1)	550(1)	825(1)
.25	140	240	275(2)	550(2)	825(2)
.5	95	170	275(3)	550(3)	825(3)
.75	80	140	275(4)	550(4)	825(4)
1.0	70	120	275(6)	550(6)	825(6)
Three-Inch with 2½"-Couplings					
500 feet	315	550	550(1)	1,100(1)	1,650(1)
1,000 feet	225	390	390(1)	780(1)	1,170(1)
.25	195	340	390(2)	780(2)	1,170(2)
.50	140	240	390(3)	780(3)	1,170(3)
.75	115	190	390(4)	780(4)	1,170(4)
1.0	95	165	390(6)	780(6)	1,170(6)

NOTE: The numbers in parentheses indicate the number of pumpers that would be required to move the associated GPM over the distance in the table. For example, it takes one supply pumper and five relay pumpers to supply 390 GPM through a single line of 3"-hose to an attack pumper one mile from the source.

Table 15.2 shows that transport distances are much longer for LDH than when only 2½"- or 3"-hose is available. Four-inch hose could supply 500 GPM one-half a mile from the source, while 5"- could transport it as much as 1½ miles if a pumper were available at the supply point. A hydrant with 50 PSI residual pressure discharging 500 GPM could provide a flow of 500 GPM over a distance of more than one-mile through 5"-hose, using hydrant pressure alone. With LDH, it is more often the amount of hose carried on initial units that limits transport distance while supplying 500 GPM, than the friction loss in the hose. While the friction loss in different types of LDH varies, this chart provides an approximation of its capabilities and is a good place to start in water supply planning. More accurate calculations can be made by actual tests of the LDH a specific department uses.

Table 15.2. Estimated Maximum Flow in GPM Using Large Diameter Hose

Distance to Source in Miles (One Way)	Number of Pumpers Required to Deliver Flow to Attack Pumper				
	HYD ONLY (50 PSI)	1 S	1S 1R	1S 2R	1S 3R
4″-LDH					
.25	425	750	1075	1325	1525
.5	300	550	750	925	1075
.75	250	425	600	750	875
1.0	225	375	525	650	750
1.5	175	300	425	525	625
2.0	150	275	375	475	550
2.5	125	250	350	425	475
5″-LDH					
.25	725	1,275	1,800	2,200	2,550
.50	525	900	1,275	1,550	1,800
.75	425	725	1,025	1,275	1,550
1.0	375	625	900	1,100	1,275
1.5	300	525	725	900	1,025
2.0	250	450	625	775	900
2.5	225	400	525	700	800

NOTE: This chart lists the supply pumper (S) and the number of relay pumpers (R) required to deliver the specified flow to an attack pumper at varying distances from the source.

TANKER TRANSPORT CAPABILITIES

If you are using tankers to transport water, estimate the continuous flow capability according to the methods suggested in Chapter 5, and calculate the transport distance. Use Table 15.3 to estimate flow when you lack accurate information for specific tankers.

Table 15.3. Continuous Flow Capabilities

Distance to Source in Miles (One Way)	Tanker Size-Usable Gallons of Water					
	1,000	1,500	2,000	2,500	3,000	3,500
.25	196	246	282	309	330	347
.5	168	216	252	279	302	320
.75	147	192	227	255	278	297
1.0	131	173	207	235	258	277
1.5	107	145	176	202	225	244
2.0	90	124	153	178	199	218
2.5	78	109	136	159	179	197
3.0	69	97	122	143	163	180
4.0	56	80	101	120	137	153
5.0	47	67	86	103	119	133
6.0	41	58	75	90	105	118
7.0	36	52	67	81	94	106
8.0	32	46	60	73	85	96
9.0	29	42	54	66	77	88
10.0	26	38	50	61	71	81

NOTE: 1. Chart assumes average fill and dump rate of 1,000 GPM.
2. Average speed of 35 MPH used for calculations.
3. An allowance of 90 seconds for maneuvering and handling hoselines is included.

According to Table 15.3, if two 2,000-gallon tankers were available, a 400-GPM flow could be maintained over a distance of one mile, but if each tanker carried 1,500 gallons of usable water (or less), it would take three of them to transport 400 GPM over a one-mile distance. On the other hand, three 2,000-gallon tankers could move 400 GPM as far as three miles. These calculations assume tankers that can dump and load at a rate of 1,000 GPM in under two minutes, and that a source that can supply at least 1,000 GPM is used.

WATER SUPPLY COVERAGE MAP

After you calculate transport distances, develop a water supply coverage map by applying the transport distance to each identified water supply point. Figure 2-7 in Chapter 2 is an example of a water supply coverage map based on a transport distance of two miles. With a transport distance of less than two miles, it takes more water supply points to provide complete coverage; for transport distances greater than two miles, fewer water supply points are needed.

TANKER COVERAGE MAP

If you have decided to use tankers to provide the base flow in certain areas, you also will need a tanker coverage map. When the same tankers respond throughout a service area, there is no need for a tanker coverage map, because the total CFC would be the same anywhere. If different tankers respond in different segments of the service area, the expected CFC will vary; then it is necessary to estimate the available fire flow for each coverage area. Use either tanker test results or Table 15.3 to derive a CFC rating for each tanker that responds in a tanker coverage area, based on the transport distance already established. Calculate the total CFC rating for a tanker coverage area by adding the individual CFC ratings for each initial response tanker.

In determining the CFC for Fire Department Water Supply under the ISO, only tankers housed within five miles of the coverage area can be used. Then, of the tankers available within the five-mile limit, only those dispatched automatically on all structure fires can be used for ISO credit, but all of them are available to increase the total water supply and fire flow. When available tankers are of different sizes, the number dispatched on an initial alarm may vary in different portions of the service area. With a tanker coverage map, similar to Figure 2-4 in Chapter 2, the WSO can determine how many tankers are needed, and how much water they will be able to supply.

SPECIFIED FLOW

When your risk analysis process shows that you need more than the base fire flow for a specific structure or target hazard, prepare a water supply preplan. Use the needed fire flow to determine the water supply points, fire hose, pumpers, and mobile water supply apparatus needed to supply it, and remember to include the mutual aid apparatus and equipment that might be needed to supply the total needed flow. Your preplan should specify: apparatus to be used; where it will come from; and what flow you anticipate will be supplied. Reach an agreement with the mutual aid departments that specifies when they will be dispatched, and what apparatus, equipment, and service they will provide. Make sure that dispatch personnel get this information so that they can send the right apparatus quickly.

AUXILIARY WATER SUPPLY EQUIPMENT

There are many areas where fire departments lack the apparatus and equipment to transport the water needed for lengthy or major emergencies, even with the help of a good mutual aid system. Transportation accidents, hazardous materials incidents, water system failures, and other serious natural or manmade emergencies could require additional water.

You will recall from earlier chapters that there are different types of vehicles available in most rural and suburban areas that can be used, with very little modification, for fire department water supply. These vehicles include tank trucks used by swimming pool companies or well drillers, transit mix trucks and commercial tank trucks. The water supply officer needs to identify any vehicles that could be used for water supply in an emergency and arrange with the owners to have them respond when needed. Any necessary adapters or special appliances should be bought or manufactured, and stored where they are ready for emergency use. The fire department and the vehicle owners agree, in writing, that the vehicle will be used, what compensation will be given, and how it will be paid. Drivers for the vehicles are identified and arrangements made to notify them in an emergency. The WSO trains the drivers, including actually loading and dumping their vehicles during a dry run to help them develop the necessary skills. Police should be notified of these arrangements to prevent any interference with the movement of these vehicles when they are used as emergency water haulers.

LOGISTICAL SUPPORT

Extended water supply operations need certain support services, of which one of the most important is fuel supply. Arrange to have a supplier send a commercial tank truck to the incident scene to refuel apparatus, even if the incident occurs during off-hours.

If you cannot arrange commercial delivery, fuel may have to be delivered to the incident scene in cans filled from a bulk storage tank at the fire station, or a service station. Either way, the WSO makes the arrangements. If you must bring fuel in cans to the fire scene, make sure there are enough cans, a supply of fuel to fill them, and a vehicle to deliver them to the incident scene.

For prolonged water supply operations, you need relief drivers and other personnel. You could need snow plows or such heavy equipment as bulldozers, backhoes, or tow trucks. Food, drinks, and other amenities are needed during lengthy operations. It is up to the water supply officer to make all arrangements, and communicate them to the dispatch center so its personnel can call for whatever assistance is required.

TRAINING

Do not omit water supply from your department's training program. Pump operators need advanced training in hydraulics and water supply techniques, and company officers need to learn the best way to set up a relay, organize a water shuttle, operate a fill site or dump site, and design an adequate water supply. Supplement formal training with dry runs based on one of your preplans. This should be a practical water supply exercise that uses the same water source, apparatus, and means of transport specified in the preplan for a structure.

Although a water supply officer generally is not responsible for setting up a training program, he or she should work closely with the department's training officer to have water supply evolution training included.

ISO RATINGS

The WSO needs to review, and be involved in, all of the different components included in the ISO's rating of the water supply. Remember, many of the aspects of the water system are not controlled by the fire department. For example, fire hydrants contribute more than 10% of the possible points awarded to the water system, so the WSO should work with the water department to see that the system's testing and maintenance program qualifies for maximum credit from the ISO.

Portions of the 50% the fire department contributes to the ISO evaluation relate to water supply as well. The water supply officer should remember that:

- Out of a possible 654 points that can be awarded for an engine company, records of the last three annual service tests on the pumpers and annual pressure tests of all fire hose account for 150;
- If a pumper carries less than 1,200 feet of 2½"- or larger hose for

supply line, the rating for the associated engine company will be reduced proportionately; and
- Tankers or large diameter hose can be used to make up for deficiencies in the hydrant system.

OPERATIONAL RESPONSIBILITIES

As soon as the Incident Commander fills the WSO role, it is up to this individual to assume full responsibility for providing the needed fire flow.

Some departments use attack pumpers with water tanks as large as 2,500 GPM; a few carry more than 1,000 gallons, but the most common size for a water tank on a pumper is 500 gallons. Even a minimal attack (two initial attack lines with a 2½"-backup) needs a dependable water supply within five minutes of arrival. The way to provide it is to establish the water supply organization as soon as possible.

When the Incident Commander appoints the WSO, he or she should specify the attack lines that will be used, and then base the estimate of the needed fire flow on the attack lines in service. Since the flow rate can be changed at the nozzle while attack crews are inside a building, out of sight of the pump operator, there is no way to determine the actual flow unless the pumper has flow meters. You do not need to know the exact flow because the WSO has to supply the attack lines with all nozzles on the maximum flow setting regardless. But, for purposes of water supply planning, the flow required for various attack lines can be estimated

Maximum Flow from Attack Lines

1½"-handlines	125 GPM
1¾"-handlines	150 GPM
2"-handlines	200 GPM
2½"-handlines	250 GPM
Master stream	350 to 1,000 GPM as specified

CONTROLLING USAGE

The WSO has to have full authority over water use, and be able to control when and where attack lines can be put into service. If the

Incident Commander tries to put more hoselines into service than the water supply operation can support, it could diminish or interrupt the flow or endanger firefighters on the attack line. It is up to the WSO not to authorize additional lines until the water supply can support them. Chapter 4 includes a method for estimating maximum flow and the types and numbers of hoselines needed to supply it. It is the WSO's job to design a water supply operation that can meet these needs and provide an adequate reserve.

SELECTING A WATER SOURCE

After you determine needed fire flow, identify, the supply source. Evaluate each potential water supply point in the service area in advance, and estimate the maximum continuous flow it will supply based on the suggestions in Chapters 6 and 7. This information always should be available to the Incident Commander and water supply officer.

Incidents in structures that were inspected should have the available water sources listed in the preplan, but if you have no preplan, refer to the water supply point inventory to find the best source to use, and pick one based on the maximum flow rate it can sustain. With a direct supply or a relay through hoselines, you want a source that can provide the maximum needed flow. For water shuttles, the source's capacity may limit the maximum CFC (as illustrated in Table 15.3) if it cannot supply a minimum of 1,000 GPM.

Of particular importance is the available flow from the fire hydrants that might be your source. Fire hydrants should be painted with a color code, as recommended in NFPA 291, *Recommended Practice for Fire Flow Testing and Marking of Hydrants*. Other than their color, most fire hydrants look alike, so unless they have been tested, documented, and marked, it is hard for a WSO and a pump operator to know whether a given hydrant supplies a maximum of 500 GPM or 2,000 GPM. Without this information, it is pointless to design a water supply operation to provide a flow of 1,500 GPM if the hydrant you are using can flow only 500 GPM.

With no potential source that will provide the needed flow, you might have to design a water supply that uses more than one source. It could be two separate operations, or two sources to supply one attack pumper.

After you select the source, designate an engine company to go to the water supply point and stand by. Specify that one of the engine companies dispatched on the initial alarm be dedicated to water supply. If the first-due station can provide two engine companies, designate the second as a water supply company. If only one engine company responds initially give the first mutual aid engine company the responsibility for water supply. If there is no previously designated water supply company, the WSO should designate one of the responding engine companies to set up the water supply, based on the mutual aid company's travel time to the water supply point, pump capacity, and the water supply equipment its pumper carries.

METHODS OF TRANSPORT

There are only two methods of delivering fire flows of 500 GPM or more in areas more than 1,000 feet from a hydrant: use hydrant pressure or a pumper to force it through fire hose, or haul it using fire department tankers. There are proponents of both methods. Many fire departments have chosen sides, and concentrated on one method or the other. Some departments use both, selecting the one that is most appropriate. The southwestern Pennsylvania New Centerville and Rural Fire Department is a good example. It operates two pumpers, each with a 1,000-gallon water tank and a load of 5″-large diameter hose (for a total of 2,500 feet) to provide fire protection to one municipality and three townships without a fire hydrant in its service area. It also operates a 2,000-gallon tanker which is backed up with a 1,500-gallon tank reserve pumper. The primary pumper has a 2,000-GPM pump; the other two have 750-GPM pumps, and the tanker has a 1,000-GPM front-mount pump and a large gravity dump outlet. With this much versatility in its equipment, a fire department's dilemma is deciding how to get the most out of the water supply capability in a particular situation.

SELECTING A TRANSPORT METHOD

Selecting a delivery method is the next step. If your needed fire flow exceeds 500 GPM, you have three choices: multiple lines of 2½″- or 3″-hose; large diameter hose; or, a water shuttle that uses tankers. Table 15.1 lists maximum flow capabilities for 2½″- and 3″-hose, Table 15.2 suggests how much water can be transported by LDH over a variety of distances, and Table 15.3 estimates the continuous flows various tankers could supply.

LARGE DIAMETER HOSE

Table 15.2 gives rough estimates of how much water could be transported through a single line of 4"- or 5"-LDH over distances ranging from .25 miles, (1,320 feet) to 2.5 miles (13,200) as listed in Column 1. Column 2 assumes that the attack pumper laid a line of LDH on the way to the incident scene and depends on a residual pressure of 50 PSI at the hydrant to deliver the water. Column 3 adds a supply pumper at the hydrant that can boost the pressure to 150 PSI, and the next three columns add relay pumpers to the layout to increase the flow. So, for example, if you need a 500-GPM flow, and the hydrant is .25 miles from the scene of the incident, 4"-hose would not supply it unless a pumper were connected to the hydrant. Five-inch hose could supply up to 725 GPM from hydrant pressure alone, and you could make the initial attack without adding a water supply pumper to the layout.

It is more difficult to provide a flow of 1,000 GPM. Five-inch hose could move 1,000 GPM nearly one-half mile with only the supply pumper at the hydrant, but 4"-hose would need a relay pumper, plus the supply pumper, to maintain the same flow.

You can increase these flows with multiple lines. When your flow reaches 1,000 GPM, add a second line of 4"-LDH instead of inserting a relay pumper in the line.

WATER SHUTTLE OPERATION

Table 15-3 lists tankers with capacities of 1,000 to 3,500 gallons of usable water that are hauling water up to ten miles. To prepare this chart, it was assumed that the fill site could supply at least 1,000 GPM, and that each tanker could load and unload at a rate of 1,000 GPM or more. An allowance of 1.5 minutes was given for maneuvering at each end of the shuttle and making or breaking hose connections. An average speed of 35 MPH (the speed the ISO uses in its evaluations) was used to calculate the continuous flow capabilities (CFC) of each tanker. However, road conditions affect actual speeds; adjust the CFC accordingly.

Based on a needed flow of 500 GPM, it would take two 2,000-gallon tankers for a supply source within .25 miles of the incident. For a distance up to five miles, it would take eight 1,500-gallon tankers, or four 3,500-gallon water haulers. For a flow of 1,000 GPM, you would need three 3,500-gallon tankers to transport it .25 miles, or eight 3,500-gallon tankers for a five-mile haul.

AVAILABILITY

With the resources outlined earlier, the New Centerville Fire Department could bring a total of 2,500 feet of 5″-LDH to an incident scene, and a 2,000-gallon tanker and 1,500 gallons of water on a pumper as well. With these, the department could use LDH or a water shuttle up to a distance of approximately one-half mile. At greater distances, New Centerville would have to wait for a mutual aid department to bring additional 5″-hose to use LDH to transport the water, but could set up its water shuttle immediately.

With one pumper for initial attack, and the other for water supply, as much as 900 GPM up to 2,500 feet from the water source could be delivered. Using its tankers to set up a water shuttle, and the second pumper for filling, the maximum anticipated flow would be slightly more than 500 GPM: 282 GPM from the 2,000-gallon tanker and 246 from the 1,500-gallon unit.

MAKING THE CHOICE

Although LDH could provide more flow up to a half-mile from the source, consider these other variables.

First, how long would it take to get the LDH in service and how many people would be needed to do it? Volunteer fire departments often are understaffed, too understaffed to set up a water supply operation.

LDH could be laid quickly with minimal connections, but no water could be supplied without hose in place, a pumper set up to draft, and a completed relay. Using tankers and a reserve pumper, the two first-in units could bring 3,500 gallons for an initial immediate attack, while the second pumper sets up to refill them.

Second, picking up a half-mile of LDH and reloading it after an incident is much more time consuming than picking up a folding tank and filling tankers in preparation for a subsequent call. Do your volunteers have enough time to do this?

Generally, LDH is best used for high flows over relatively short distances. A well-organized water shuttle can provide as much as 2,000 GPM, but use tankers to provide flows up to 500 GPM over long distances with minimal delay. The key to either type of supply is to have the right equipment and apparatus, and a plan to use it efficiently.

MUTUAL AID

These examples considered only the apparatus and equipment included in an automatic mutual aid system because most water supply operations simply are beyond the capabilities of a single engine company or fire department. If the water supply is remote, and the needed flow is more than 500 GPM, you need a coordinated approach with additional alarms to provide the apparatus needed for transport.

COMBINED OPERATIONS

A combined operation may be the solution for providing a flow that exceeds 1,000 GPM, for example, supplying 500 GPM by a relay or LDH, and 500 GPM by water shuttle.

Although a relay that uses LDH could probably stand alone, a water shuttle frequently involves using LDH at the dump site to transport the water from the nurse tanker or portable tank to the attack pumper. LDH is a way to maintain a fill rate of 1,000 GPM if tanker loading stations are more than 100 feet from the water source.

WATER SUPPLY DESIGN AND SETUP

After the WSO determines needed flow, water sources, and method of transport, he or she designs and operates the water supply. In the event of multiple relays, water shuttles, or both, or, for a combined operation, individual relay control officers or water shuttle control officers are assigned for each. In these situations, the WSO assumes overall responsibility for water consumption and supply; individual relay control or water shuttle control officers exercise operational control over individual segments of the operation. For single-water supply situations, the WSO has operational control directly over the relay or water shuttle.

The water supply officer:

- Determines needed fire flow;

- Matches the supply lines and the tanker CFC to the requirements of the attack lines;
- Assigns the water supply apparatus and equipment needed to transport the flow from the source to the incident;
- Calls for mutual aid units, if needed, to supplement the available apparatus, hoselines, and equipment as required; and
  (Include the estimated time of arrival (ETA) of mutual aid apparatus when you determine the number of attack lines that can be used safely, and dispatch enough mutual aid apparatus to provide a reserve pumper and tanker in case of mechanical difficulties or equipment failure.)
- Staffs the operation with experienced pump operators for water supply apparatus, fill site and dump site officers where required, and a traffic control organization.

OPERATIONAL CONTROL

The key to an efficient and reliable water supply operation is a water supply officer who:

- Is in control, and known to all units participating in the water supply operation;
- Communicates with all participants in the water supply operation; (Where different radio frequencies are used, distribute portable units on the primary frequency to mutual aid apparatus.)
- Maintains control of water usage. (The WSO keeps the incident commander apprised of the progress being made in establishing the water supply, any problems that are encountered, and the status of the water supply);
- Monitors all water sources by checking the residual pressure at fire hydrants, the level in storage tanks in central water systems, and the water level in ponds or other sources frequently;
- Calls for water system personnel to provide maximum production when an extended operation is anticipated;
- Prepares to provide a supplemental supply if increased flow is required or the supply of water in storage is exhausted; and
- Provides fuel, food and water for operators, and replaces personnel when needed.

WATER SUPPLY PLANNING

Water supply planning is the most important part of the water supply officer's job. A designated departmental position must delineate plans, SOPs, etc., for anyone who could be assigned the WSO job during an emergency.

He or she must consider many variables in selecting a method of water supply. For example, when tankers are used to operate a water shuttle, tanker construction, maximum flow at the fill site, storage at the dump site, road conditions, and other variables affect the actual CFC of individual tankers and specific water shuttles. The only way to determine the capabilities of mobile water supply apparatus is to test it thoroughly, implement a development plan for potential water sources, and conduct training exercises to practice hauling water under controlled conditions.

There are not as many variables to consider when using LDH to transport water, but different types of hose do have different characteristics. Construction methods, materials used, and actual hose sizes vary widely among manufacturers. Plan accurately by conducting tests of the LDH that is carried on the apparatus: flow water and record the maximum achievable flows.

Water supply planning is essential to providing an acceptable level of fire protection in a community. Planning is what makes it possible to provide adequate flows to control fires in structures before excessive damage occurs. More important, planning ensures that the supply to attack lines will not be interrupted when the crew on the hoseline needs the water for protection.

Glossary

FIRE SERVICE TERMINOLOGY VARIES by department type, geographic area of the country, and agency involved. This glossary was written for this book, not to establish a standard for the fire service, but to clarify the author's meaning in using certain words. In general, the terms used in this book conform to the meanings used by the ISO in its 1980 rating guide and to the terms used in the NFPA standards referenced in the text.

Acceptance Test-Test(s) conducted at the time of or after delivery of new fire apparatus to verify that it meets the purchaser's specifications before it is accepted by the purchaser.

Application Rate-The amount of water in Gallons Per Minute (GPM) applied to a fire.

Atmospheric Pressure-The weight of air, 14.7 PSI at sea level, decreasing approximately .5 PSI with every 1,000 feet of increased elevation.

Attack Line-Hoseline used to apply water to a fire.

Attack Pump-Pump installed on fire apparatus with a minimum capacity of delivering 250 GPM at a net pump pressure of 150 PSI.

Attack Pumper-Pumper used to provide needed fire flow at the required pressure for attack lines.

Authority Having Jurisdiction-The organization, official governmental entity or individual that has the authority to approve equipment, procedures, or activities within certain boundaries.

Automatic Mutual Aid-Apparatus/personnel/equipment from area fire departments dispatched automatically and simultaneously with the fire department that has the primary responsibility for fire protection, generally in accordance with a written agreement.

Glossary

Automatic Nozzle-Nozzle for fire attack designed to maintain a constant pressure at the discharge by changing flow to regulate the friction loss in the hoseline.

Automatic Pressure Control Device-Device used to limit pressure changes on the intake or discharge of a fire pump. A relief valve or a pressure governor can be used for this purpose.

Automatic Relay-A water-moving relay that uses automatic pressure control devices to maintain needed pressure during flow changes in accordance with the procedures suggested in Chapter 8: Relay Operations, of this book.

Automatic Suction Valve-Valve designed for use on the intake of a pumper drafting from a portable tank that will transfer the supply to the fire pump from the tank on the apparatus to the portable tank automatically with pressure changes at the intake to the pump.

Auxiliary Water Supply Equipment-Commercial vehicles, generally privately owned, capable of hauling liquids, but not constructed for fire department use.

AWWA Manual M17- Manual of Water Supply Practices-Recommended practices and procedures for the operation of a water system, including fire hydrants, published by the American Waterworks Association.

Back Pressure-Pressure created by the weight of water when the water level in a system or conduit is higher than the point of measurement. (To overcome back pressure, an attack pumper must add 5 PSI to the discharge pressure for each floor in a building below and including the floor of anticipated fire attack. When supplying water from the source to an attack pumper over uneven terrain, 5 PSI is lost for each ten feet of elevation difference between the source and the attack pumper.)

Backup Attack Lines-Hoselines used to provide additional flow when initial attack lines cannot supply enough water for the volume of fire. Backup lines generally are two-inch or larger hoselines.

Base Fire Flow-Amount of water in Gallons Per Minute required for initial attack for fires in typical structures within a fire management area.

Booster Line-Rubber hose, ¾" or 1", generally carried on a hose reel (maximum flow of less than 60 GPM).

Booster Pump-Pump with a capacity lower than 500 GPM, mounted on a piece of fire apparatus, and generally driven by a PTO or separate engine.

Cavitation-The condition that develops inside a fire pump when more water is discharged than can be supplied from the source, causing fluctuations in flow. Cavitation can damage a pump over time.

Central Water System-Water supplied to an area from a common source through a network of pipes.

Clappered Siamese-A fitting consisting of two intakes with female couplings supplying one discharge outlet with a male coupling. Each intake has a check valve that is held closed when the pressure on the discharge side of the siamese is greater than the incoming pressure from the supply line.

Class A or Standard Pumper-Pumper constructed according to the provisions of NFPA 1901, *Standard for Pumper Fire Apparatus*, and certified to supply its rated capacity from draft with no more than a ten-foot lift while maintaining a net pump pressure of at least 150 PSI.

Closed Relay-A method of water transport that uses a pumper receiving water from a source and supplies the water through hose and a direct connection through any needed relay pumpers directly to the intake connection of an attack pumper.

Community Fire Defense, Challenges and Solutions-A National Fire Academy Train-the-Trainer course on methods of analyzing community fire protection needs and presenting alternate means of providing for them.

Community Impact-A measure of the impact on a community of a structure lost to fire, over and above the building's monetary value.

Comprehensive Water and Sewer Plan-A master plan for extending and constructing water and sewer systems throughout a political subdivision.

Constant Flow Relay-A method of operating a fire department relay by dumping water to maintain a constant flow and minimize pressure changes. (Described in Chapter 8: Relay Operations, of this book.)

Continuous Flow Capability-A measure of a tanker's (Mobile Water Supply Apparatus) efficiency and a means of estimating the amount of water it can supply on a continuous flow basis when it is operating in a well-organized water shuttle.

Coverage Map-A map with lines indicating the area that falls within a specified distance of a fire station, water source, or apparatus that will respond.

Defensive Attack-An attack approach of controlling fire spread when the resources available to extinguish it are insufficient.

Delivery Rate-The rate in GPM of water being supplied to an incident scene.

Diamond Pattern-The pattern created when one corner of a portable tank is pointed toward the suction inlet of a drafting pumper and the sides are at a 45-degree angle, during a water shuttle.

Direct Hoselines-Hoselines used to supply water to a pumper directly from a fire hydrant using the system's residual pressure instead of pump pressure to force the water through the supply lines.

Discharge-A descriptive term to indicate that an outlet, gauge, or hoseline is connected to the pressurized side of a fire pump.

Discharge Tube-Tube used to direct water flow from the discharge outlet of a fire pump into a portable tank during a water shuttle evolution.

Draft-The process of taking water from a nonpressurized source by removing enough air from the pump to create a pressure differential. The differential allows atmospheric pressure to force water from the source through the suction hose into the fire pump.

Drafting Hydrant-Hydrant connected to a water source with less than 20 PSI static pressure. The pumper must draft from the hydrant and develop enough vacuum to overcome the friction loss in the suction hose and pipe to the source as well as any head pressure.

Drafting Supply-A water source that a pumper can use to draft and supply needed flow.

Drop Tank-Portable tank used for water storage at a dump site to sustain a water shuttle operation.

Glossary

Dry Hydrant-Drafting hydrant located with the pumper connection above the water's surface. Hydrant barrel remains partially dry until a pumper connects to it and creates enough vacuum to overcome the lift.

Dry Run-Test of a prefire plan during which planned procedures are followed to determine strengths, weaknesses and usability.

Dump Extender-Device installed on a tanker's dump outlet to simplify dumping into a portable tank.

Dump Line-Hoseline connected to a discharge outlet on a mobile water supply apparatus that can be used to pump its load into a portable tank or to an unused outlet on the pumper at the terminal end of a "constant flow" relay to get rid of excess water.

Dump Rate-The average GPM flow from the dump on a tanker. Determine dump rate in GPM by dividing the amount of usable water a tanker carries by the dump time.

Dump Site-The area where tankers unload during a water shuttle.

Dump Time-Dump time begins in tanker testing when water starts to flow from the dump, and ends when usable water has been discharged. In water shuttle design, overall dump time begins when a tanker arrives at the dump site, and ends when it leaves to go to the fill site for another load.

Dumping Station-A location where a tanker can discharge its load during a water shuttle.

Dynamic Pressure Losses-Pressure losses related to water movement, generally caused by friction as water moves through hoselines and appliances, that change with flow rate.

Elevated Storage Tank-Storage tank in a water system constructed on a superstructure raised high enough above the ground to maintain the desired pressure in the system.

Elevation-Height above sea level at a particular location. Differences in elevation between the ends of a hoseline create a pressure difference of .434 PSI per foot (an allowance of .5 PSI per foot or 5 PSI for each ten-foot difference in elevation is adequate for most field calculations).

Engine-A term frequently used to refer to a fire department pumper or the motor that powers a vehicle.

Engine Company-A pumper, officer, and enough firefighters to launch an attack on a fire. NFPA 1500 recommends a minimum four-person engine company.

Engine Pressure-The pressure developed by a pumper.

Evaluator-An ISO Field Representative who conducts a survey of an area's fire protection capabilities and assigns a Public Protection Classification Rating.

FAWR-A vehicle's front axle weight rating as determined by the manufacturer; a measure of the weight a vehicle will carry safely.

Fiberglass Tank-A lightweight, durable water tank constructed to reduce the total weight of a mobile water supply apparatus.

Fill Line-Hoseline used for filling tankers.

Fill Rate-Average GPM flow into a tanker. (Calculate fill rate by dividing the usable water the tanker carries by the fill time.)

Fill Site The area where tankers are loaded during a water shuttle.

Fill Time-Fill time in tanker testing begins when water starts to flow into the tank, and ends when water flows out of the overflow or vent. In water shuttle design, it begins when the tanker arrives at the fill site, and ends when it leaves to return to the dump site with a load of water.

Fire Department Water Supply (FDWS)-Water delivered by a fire department to an area more than 1,000 feet from nearest available hydrant. To receive ISO credit, a fire department must supply at least 250 GPM initially and continue or increase the flow for the fire's duration based on Needed Fire Flow, using either hoselines or tankers to transport the water.

Fire Fee-Fee levied on property owners in special-purpose fire districts, or political subdivisions to pay for fire protection services.

Fire Flow Worksheet-Computer-generated worksheet used to track available water supply, compare it to needed fire flow to determine water supply needs, and to develop water supply preplans.

Fire Hydrant-Device supplied by a central water system and designed to deliver at least 500 GPM with a pressure of at least 20 PSI at the hydrant.

Fire Management Area-An area that presents similar fire problems, including the type of structures, needed fire flow, and available water supply that can be served by the same fire station or stations.

Fire Pump-Pump installed on fire apparatus with a minimum capability of delivering 750 GPM at a net pump pressure of 150 PSI.

Fire Tax-Property tax paid for the purpose of receiving fire protection. Can be levied across a political subdivision, or be confined to special-purpose districts.

Fire Wall-A wall of sufficient durability and stability to withstand the effects of the most severe anticipated fire exposure. Openings, if allowed, must be protected. (*Fire Protection Handbook*, 17 Ed., NFPA, 1992)

Fireground Hydraulics-Methods used to estimate losses incurred and pressures needed to maintain an effective water supply on the fireground.

First Alarm Assignment-Normal complement of apparatus and personnel dispatched to a structure or within an area on receipt of an alarm.

Flexible Tank-Portable water storage tank constructed of fabric without a rigid framework and with no fixed shape. (Can be stored in a very small area when collapsed, and has a floating collar that contains the water inside the tank as it is filled.)

Friction Loss-Pressure loss caused by the friction of water against a waterway's surface as water moves through a conduit.

Front-Mount Pump-Fire pump mounted in front of a pumper's radiator and driven from a Power Take-Off (PTO) unit mounted on the front of the crankshaft. (Can be used to develop pressure and supply water whether the vehicle is stationary or in motion.)

Frost Line-The limit of penetration of soil by frost during a normal winter. Water stored below the frost line is safe from freezing.

Fuel Loading-A measure of the amount of fuel contained in an enclosed space. An estimate of the fuel loading provides a means of anticipating the maximum volume of fire to be expected and the amount of water needed for extinguishment.

Gated Wye-A fitting with one intake, equipped with a female coupling, and two discharge outlets, both with male threads. Each discharge outlet has a valve, and can be individually controlled.

General Fund-Monies generally used by local governments to pay for the cost of the various services they provide. Different revenue sources comprise this fund, and expenditures cover a wide variety of activities.

General Purpose Tanker-Multi-purpose tanker that can be used for water supply in a water shuttle, as a nurse on the fire scene, or to make a fire attack.

Governor-Mechanical device used to control engine speed.

Gravity Dump-An outlet, at least 4½″ in diameter, that allows a tanker to dump its load into a portable tank without the use of a pump.

Gridded Water Distribution System-A water system of interconnected water mains, designed to supply the maximum amount of water under varying usage conditions by providing alternate routes for water movement through the system.

Ground Level Storage Tank-A water system storage tank constructed at ground level, that uses high-service pumps to supply water to the system or to transfer the contents of the storage tank to elevated tanks used to maintain system pressure.

GVWR-Gross Vehicle Weight Rating assigned by the manufacturer and a measure of the maximum weight a vehicle can carry safely as it has been constructed.

Hand Method-A memory jogger to estimate expected friction loss in a 100-foot section of fire hose. (See Chapter 4: The Use of Fire Hose in a Water Delivery System, of this book.)

Head Pressure-Pressure created by water weight. (Figure 7-3, Chapter 7: Suction Supply as a Water Source, shows that a column of water 10 feet high, frequently referred to as 10 feet of head pressure, would create a pressure of 4.34 PSIg at its base.) Also known as back pressure.

Heavy Attack Line-See Backup Attack Line.

High-Capacity Fill Site-A fill site operation that can sustain a continuous flow greater than 500 GPM in a water shuttle evolution.

High-Capacity Pumper-Pumper with a fire pump larger than 1,000 GPM.

High-Capacity Pumper Tanker-Pumper tanker that carries more than 1,000 gallons of usable water and has a fire pump with a capacity greater than 1,000 GPM.

High-Capacity Tanker-Tanker that carries more than 1,500 gallons of usable water.

High-Capacity Water Tank-Water tank on a Standard pumper that carries 1,000 gallons or more of water. A pumper with a high-capacity tank also can be referred to as a pumper tanker.

High-Capacity Water Shuttle-Water shuttle evolution designed to produce a continuous flow greater than 500 GPM.

High-Pressure Storage-Water storage in a central system designed to provide continuous pressure to the distribution system. Water can be stored in elevated tanks, standpipes, high-elevation reservoirs, or pressurized tanks.

Hose handlers-Individuals designated to make and break tanker hose connections at fill and dump sites in a water shuttle.

Hoseholder-Device designed to hold hose in position while water is discharged under pressure into a portable tank.

Hose Tests-Annual pressure tests conducted to verify fire hose condition. (Test results become part of the history for each individual section of hose.)

Hydrant Gate Valve-A gate valve for use on the outlet of a fire hydrant to provide individual flow controlfrom outlets without opening and closing the main hydrant valve.

Hydrant System-A fire protection water system, including hydrants, distribution network, supply source, and storage, to maintain required pressure when water is withdrawn from a hydrant.

Hydraulics-Study of the use and movement of water.

Initial Attack Lines-Initial attack hoselines, typically preconnected lines of 1½"- or 1¾"-hose with a variable flow, variable pattern nozzle.

Initial Water Supply-Water carried by first-alarm units.

Insurance Services Office (ISO)-An organization established by the insurance industry to evaluate community risk level and provide guidance in the establishment of fire insurance rates.

Intake Relief Valve-Device to limit incoming pressure to a fire pump. (When incoming pressure exceeds the preset value, the valve opens and allows water to discharge outside the pump; this limits the pressure increase by increasing the friction loss in the supply line.)

ISO Rating Schedule-Guidelines used by ISO evaluators to review community fire suppression facilities to develop a Public Protection Classification for fire insurance rating purposes.

Jet Dump-Small stream of water from the fire pump used to accelerate water movement through the dump outlet and increase dump rate from a tanker.

Ladder Company-Aerial apparatus with officer and crew who perform ladder or truck company tasks on the fireground.

Lateral Pipe-Portion of pipe that extends from a dry hydrant to the strainer and the source of water.

Lift-Distance from water's surface to the eye of the impeller of a pump drafting from a static source of water.

Light Attack Line-Handline used for initial attack. (See Initial Attack Line.)

Loading Station-The location where tankers will load at a fill site during a water shuttle.

Low-Pressure Storage-Water stored at ground level under atmospheric pressure only. High-service pumps must be used to transfer this water to the high-pressure storage system or make it available for fire protection purposes by direct pressure to the system.

Manifold-Hose fitting with one intake, either with female threads or a sexless coupling on the inlet, and as many as five outlets that vary in size and connections. Each outlet should have a valve and be controlled individually.

Mobile Water Supply Apparatus-Fire department apparatus equipped with water tanks of 1,000 gallons or greater capacities.

Mutual Aid-A formal agreement made between fire departments to assist each other on an automatic (both departments dispatched automatically on all alarms) basis, or on an as-needed basis, depending on the terms of the agreement and the needs of the area.

Needed Fire Flow-Estimated amount of water needed to effect fire control in a structure in a reasonable amount of time.

Negative Pressure-Pressure less than atmospheric pressure, generally measured in inches of vacuum.

Net Pump Pressure-Difference between a pump's intake and discharge pressures. Net pump pressure is a measure of the amount of effort a pump must exert to attain required pressure at the discharge. (When it operates from draft, a pump must supply the negative pressure to lift the water into the pump as well as the positive pressure on the discharge; the net pump pressure will always exceed the reading on the discharge gauge. When water is received under pressure on the intake, net pump pressure will always be less than the reading on the discharge gauge.)

NFPA 1001, *Standard for Fire fighter Professional Qualifications*-Skills, knowledge and abilities needed for firefighter certification.

NFPA 1231, *Standard on Water Supplies for Suburban and Rural Fire Fighting*-Water supply practices for suburban and rural areas, including determining needs and suggested methods to meet them.

NFPA 1410, *A Training Standard on Initial Fire Attack*-Firefighter initial attack training standards. Sets initial attack objectives, and provides guidelines for determining water needed to support a fire attack of a suggested magnitude, and how soon after arrival it will be needed.

NFPA 1500, *Standard on Fire Department Occupational Safety and Health Program*-Guidelines for firefighter safety and health. Includes hazards to eliminate, and a recommended organizational structure for fire department health and safety.

NFPA 1901, *Standard for Pumper Fire Apparatus*-Minimum requirements for fire department pumper construction.

NFPA 1903, *Standard for Mobile Water Supplies*-Minimum requirements for constructing fire service mobile water supply apparatus.

NFPA 291, *Recommended Practice for Fire Flow Testing and Marking of Hydrants*—Routine maintenance procedures and service testing of fire hydrants.

Nurse Tanker-A fire department tanker that remains at a dump site during water shuttle operations. Tank is used as a reservoir for shuttle tankers to pump into to maintain a reserve water supply.

Offensive Attack-Initial attack intended to effect fire control quickly and extinguish fire before it spreads to adjacent exposures.

Open Relay-A method of water transport during which a pumper receives water from a source and supplies it through fire hose and any needed relay pumpers into a portable tank located near an attack pumper. The attack pumper drafts from the tank to provide the water supply needed to support attack lines.

Overflow-Opening in the tank to permit water to escape when it is filled.

Overhaul-A post-fire process of checking for areas of possible rekindle.

Parallel Tank-Portable tank positioned approximately 18 to 24 inches from a drafting pumper with one side parallel to the side of the pumper.

Petroleum Type Coupling-Quick-connect coupling for fire hose coupling and uncoupling. Two levers are operated to make the connection. (This is similar to the method used to connect fuel oil trucks to their external lines.)

Pneumatic Tank (Hydro-Pneumatic)-Ground-level tank designed to be pressurized by the pumps to maintain adequate pressure in a central water system.

Polypropylene Tank-Lightweight water tank, generally with welded seams, used on fire department pumpers and mobile water supply apparatus. Polypropylene does not deteriorate over time as is the tendency of metal tanks.

Portable Hydrant-Manifold supplied by Large Diameter Hose with multiple individually controlled discharge outlets to provide supply water to multiple pumpers or hoselines. Generally has a pressure gauge to determine amount of pressure available to supply hoselines.

Portable Tank-Collapsible tank carried on a pumper or tanker, and used on scene as a reservoir for tanker water. The water supplies attack pumpers.

Positive Pressure-Pressure higher than atmospheric. A pressure gauge normally reads positive pressure, and 0 PSIg actually represents a reading of 14.7 PSI at sea level.

Power Take-Off (PTO) Pump-Pump driven by a gearing arrangement from the driveline of a vehicle. A PTO pump generally permits pump and roll operation.

Power Transfer-Device designed to use a section of hard suction hose or other conduit, and a 1½"-hoseline from a discharge outlet on the pump, to transfer water from one portable tank to another during a high-capacity water shuttle.

Preplan-A plan developed after an analysis of the needed fire flow and special hazards in a fire department's response area. The preplan should specify water adequate to support an offensive attack and effect fire control.

Pressure Governor-Device used to limit pressure increases in a fire pump by automatically changing engine speed to adjust the pressure.

Pressurized Dump Lines-Hoselines used at a dump site during water shuttle operations for any mobile water supply apparatus equipped with a fire pump to deliver its load under pressure to the portable tank or nurse tanker.

Pressurized Tank-Water system storage tank designed to allow high-service pumps or an air compressor to provide the pressure needed to supply the system.

Primary Risk-In preplan development, a building considered to be the primary risk; nearby exposures are considered secondary risks.

Primer-Device that removes enough air from a centrifugal pump to create a pressure differential inside the suction hose and pump housing to enable atmospheric pressure to force water into the pump until it is primed.

Public Protection Classification (PPC)-Insurance company rating of 1 to 10 (1 = best, 10 = worst) used to set community property fire insurance rates.

Pumper Outlet (Hydrant)-Fire hydrant outlet more than 2½" in diameter.

Pumper-Tanker-Standard pumper with a water tank capacity of at least 1,000 gallons.

Quick-Connect Coupling-Fire hose coupling that can be connected and released with one motion without the use of standard fire hose threaded couplings. These can be camlock, quarter-turn or compression type.

RAWR-The rear axle weight rating, set by manufacturers, that specifies the amount of weight the rear axle on a vehicle can carry safely.

Reflex Time-Time from receipt of alarm at alarm center until first engine company leaves to respond.

Relay-Inserting pumpers into a hoseline between the supply pumper at the source and the attack pumper to compensate for losses experienced and ensure adequate fire flow.

Relay Control Officer-Officer designated in control of a relay under the command of the water supply officer.

Relief Valve-Automatic pressure control device connected to discharge side of a pump to prevent excessive changes in discharge pressure with changes in flow through the pump. When discharge pressure increases, relief valve opens to allow enough water to circulate around pump to return pressure to original reading.

Remote Unloading Station-An unloading station set up some distance from the storage tanks for tankers that can pump off their water through hoselines.

Reserve Engine-Pumper held in reserve to replace an in-service unit not available for any reason.

Reserve Tank-Portable tank used at a dump site for high-capacity water shuttle operations. The tank holds a reserve water supply to sustain needed flow when no tankers are dumping. The reserve tank is not used by drafting water supply pumpers. Water must be transferred into a supply tank before it is usable.

Reservoir-Water Storage facility, generally at ground level or underground, at atmospheric pressure. A sufficiently elevated reservoir can be used for high-pressure storage. At lower elevations, it is used for low-pressure storage; high-service pumps are needed to get the water into the system.

Residual Pressure-Pressure remaining at a hydrant with water flowing through supply lines to the attack pumper.

Response Time-Time from receipt of an alarm at alarm center until first engine company arrives on scene.

Rigid Frame Tank-A folding, transportable tank with a rigid metal frame that retains its size and shape wherever it is set up.

Risk Analysis-The process of determining needed fire flow, identifying special hazards, selecting water sources, arranging necessary apparatus and personnel, and developing a plan of attack for a specific structure or facility.

Service Company-A vehicle that carries specialized equipment, an officer and crew sufficient to perform ladder company tasks. A service company responds into areas not requiring a ladder company.

Service Tests-Tests that verify whether a pumper can perform as it did when originally constructed. Perform service tests annually, according to the procedures and requirements listed in Chapter 5: Testing Water Supply Apparatus, of this book and NFPA 1911. List test results in unit's permanent maintenance records.

Siamese-Hose fitting with two inlets, each equipped with female couplings, and one outlet equipped with a male coupling.

Siphon-Device designed to transfer water between portable tanks used at a dump site during water shuttle operations. Uses force of gravity to move water and will maintain the same level in both tanks.

Soil Conservation Service (SCS)-A branch of the United States Department of Agriculture concerned with soil conservation and erosion control.

Special Risk-A structure whose life safety concerns, exposure potential, high fuel load, hazardous materials storage or other specific concerns deem it more hazardous than others encountered in a fire management area.

Standpipe Tank-A tank set on a foundation at ground level, but tall enough to create the pressure needed to operate a system. Generally cylindrical, it is the same diameter and shape from bottom to top.

Static Pressure-Pressure in a system before any water movement occurs.

Static Pressure Loss-Pressure losses caused by changes in elevation, atmospheric pressure, or other factors not affected by the amount of flow through a system.

Static Source-A ground level water source unusable without a drafting pumper.

Subsequent Alarm-A call for additional apparatus and personnel needed after initial response units arrive.

Suction Gated Wye-Gated wye with the intake side of the fitting equipped with a female coupling that fits a pumper's large intake connection.

Suction Siamese-Siamese fitting with the discharge side equipped with a female coupling that will fit a pumper's large intake connection.

Suction Supply-Use of a pumper to draft from a static source to provide needed water supply.

Sump-Area extending below the bottom of the tank to enable more water to be removed by a pump before it goes into cavitation and to provide a place for dirt and debris to accumulate where it can be removed easily.

Supplemental Suction Supply-Use of a pumper at draft to supplement supply where a water system is unable to supply needed flow.

Supply Line-Hoseline used for water transport to provide needed flow to an attack pumper. Hoselines connected to the intake connections on an attack pumper are supply lines; those connected to the discharge side of the pump generally are attack lines.

Supply Tank-Tank from which a pumper drafts during a high-capacity water shuttle evolution that uses multiple portable tanks at the dump site.

Supply Works Capacity-The ability of a supply source, treatment facilities, pumps, and water lines to supplement water in storage during an evaluation of a water system's ability to provide needed fire flow.

Target Hazard-Structure or facility located in a fire management area that requires a fire flow significantly higher than base flow, as well as additional apparatus, equipment, and manpower.

Glossary

Tender-Refers to fire department tankers in areas where the term tanker describes airplanes used to drop water on forest fires or wildland fires.

Total Water Supply (TWS)-Total water supply needed for a fully involved structure, calculated using the procedures recommended in NFPA 1231, and described in Chapter 3: Risk Analysis in Fire Protection Planning, of this book.

Transfer Pump-Pump installed on fire apparatus with a minimum capability of delivering 250 GPM at a net pump pressure of 50 PSI.

Transport-Process of moving water from a source to an attack pumper.

Transport Distance-Distance a fire department can move needed fire flow from a water source using the hose, pumpers, and tankers included in the first alarm assignment.

Travel Time-Roundtrip travel time of a tanker between a fill site and dump site during a water shuttle. Travel time is determined by distance, road conditions, and the tankers being used.

Two-Piece Engine Company-A standard operating practice of using one piece known as the "Hose Wagon," to lay hose and proceed to the fire, while the second piece, known as the "engine," connects to the hydrant and provides pressure needed to supply hoselines.

Unloading Station-The location where tankers will be unloaded at a dumpsite during a water shuttle.

Vacuum Tanker-Fire department tanker that uses a pressure vessel for a tank. The vacuum pump on the apparatus removes the air from the tank, creating a vacuum that allows it to use a static source, with hard suction hose for filling, without a pumper needed to load it.

Vent-An opening in a water tank to permit the free movement of air in and out of the tank as it is filled and unloaded.

Waiting Time-Amount of time tankers wait to load or unload during a water shuttle.

Water Delivery System-A complete water delivery system, including a water source, pumpers to move the water, and a means of transporting water to the attack pumper to supply attack lines.

Water Department-Agency responsible for maintaining and operating a central water system.

Water Fill Point-Water source identified, developed, and included in a fire management area's water supply plan where tankers can be filled during water shuttle operations.

Water Hammer-A hydraulic shock transmitted to pump and hoselines upon the sudden closing of a discharge valve or nozzle.

Water Shuttle-System of using transport tankers to deliver a constant flow from a water source to an attack pumper.

Water Shuttle Control Officer-Officer designated in control of a water shuttle under the command of the water supply officer.

Water Supply Alarm-Apparatus and equipment to provide a specified amount of water to a specific location, or a general area, within a fire management area. Specific apparatus and equipment are assigned in advance based on needed flow; this information is provided to the dispatch center.

Water Supply Point-An identified and developed water source for which a signed agreement between the property owner and the fire department exists, allowing the source to be used for supplying hoselines or filling tankers.

Water Supply Pumper-Pumper used to supply an attack pumper. During a relay, a pumper that takes water from a source and delivers it to a relay pumper typically is designated as the water supply pumper.

Water Supply Tanker-Mobile water supply apparatus designed to deliver water to the scene of an incident and categorized as:

WST-P-Equipped with a fire pump.

WST-A-Equipped with an attack pump.

WST-T-Equipped with a transfer pump.

WST-N-No pump installed.

Index

A

Acceptance Test, 376, 400
Anti-swash partitions, 358
Anti-swirl device, 324, 368
Apparatus, 40
Apparatus Evaluation, 45
Apparatus Replacement Schedule, 43
Application Rate, 58
Atmospheric Pressure, 92, 173
Attack Lines, 68, 72
Attack Pump, 107
Attack Pumper, 3, 83, 84, 386
Authority with Jurisdiction, 29, 315, 396
Automatic Mutual Aid, 14, 31, 50, 239, 389
Automatic Pressure Control Device, 101, 228, 233
Automatic Nozzles, 70
Automatic Relay, 232
Automatic Suction Valve, 44, 280, 329

Auxiliary Water Supply Apparatus, 26, 340, 406
AWWA Manual M17, 158

B

Back Pressure, 97, 213
Baffles, 358
Ballast, 121, 359
Base Fire Flow, 32
Booster Line, 68, 313
Booster Pump, 107
Budget, 30
Building Codes, 6, 396
Building Inspection, 396

C

Capital Budget, 50
Capital Improvement Program, 30, 38
Cavitation 86, 122, 176
Central Water System, 125, 395

Check Valves, 374
Clappered Siamese, 18, 248, 264, 321, 381
Closed Relay, 220
Color Coded Hydrants, 155
Combination Apparatus, 83, 345
Combined Operation–High Capacity Water Shuttle, 331
Community Fire Defense, Challenges and Solutions, 32, 53
Community Impact, 51
Comprehensive Fire Protection Plan, 52
Comprehensive Water and Sewer Plan, 395
Connection Time, 263
Constant Flow Relay, 230
Construction Classification, 56
Continuous Flow Capability, 45, 104, 115, 124, 243, 250, 391, 405
Contributions, 52
Controlling Water Usage, 262
Conversion from commercial to MWS apparatus, 109, 344
Coverage Maps, 34

D

Dead End Water Mains, 135
Delivery Rate, 57
Diamond Position of Portable Tanks, 275
Diffuser, 139, 283, 342
Direct Fill Inlet, 114
Direct Fill Lines, 296, 308
Direct Hose Lines, 382
Direct Lay from Hydrant, 15
Direct Supply from Tankers, 4, 244, 264, 387
Discharge, 113
Discharge Tube, 285
Diverter, 139
Draft, 96, 206, 267, 274, 365

Drafting Hydrant, 185
Drafting Supply, 94, 306, 318
Drop Tank, 44, 248
Dry Hydrant, 188, 196, 199, 203
Dry Run, 13, 48, 66, 261, 406
Dump Extender, 285, 371
Dump line, 224
Dump Outlet, 122
Dump Rate, 122
Dump Site, 240, 249, 263, 288, 291
Dump Time, 114
Duration of Flow, 26, 165, 167
Dynamic Pressure Losses, 213

E

Electronic Flow Meter, 113, 155
Elevated Storage Tank, 128
Elevation, 94, 213
Engine, 104, 110, 376
Engine Company, 11, 40, 389
Engine Pressure, 86, 94, 214, 221
Equipment, 43
Evaluator, 32
Evaluation, 32, 83, 102
Exposure Factor, 56

F

FAWR, 108, 110
Fiberglass Tank, 356
Fill Line, 114, 300, 302
Fill Pumper, 294, 309
Fill Rate, 293, 306, 320, 357
Fill Site, 240, 293, 304, 310, 315
Fill Site Officer, 310, 312, 315
Fill Time, 114
Fire Department Water Supply (FDWS), 13, 66, 104, 188, 378, 386, 391
Fire Fee, 51
Fire Flow Worksheet, 63
Fire Management Area, 10, 28, 32, 57, 342, 379, 388

Fire Pump, 107
Fire Station Coverage Maps, 34
Fire Tax, 51
Fire Wall, 59
Fireground hydraulics, 71, 75
Fireground Officers, 241
First Alarm Assignment, 63, 379, 389
Flexible Suction Hose, 195
Flexible Tank, 270
Floating Strainer, 44, 275, 277, 278, 306
Flow Tests, 86, 88, 90, 113, 167
Four Way Hydrant Valve, 18, 78, 380
Friction Loss, 73, 77, 175, 213
Friction Loss in Attack Lines, 69
Front Mount Pump, 365
Fuel Loading, 385
Full Load, 120
Fund Raising, 52
Funding Fire Departments, 50

G

Gated Suction Connection, 280, 386
Gated Wye, 222, 266, 296, 302, 308, 310, 387
Gauges, 267
General Fund, 51
General Purpose Tanker, 107, 348
Governor, 94
Gravity Dump, 288, 348, 359, 367
Gridded Water Distribution System, 135
Ground Level Storage Tank, 131
Guidelines for Planning, 28
GVWR, 108, 352, 375

H

Hand Method, 75
Handling Time, 362
HAZMAT, 7

Head Pressure, 130, 212
Heavy Attack Line, 384
High Capacity Dump Site, 323, 332, 388
High Capacity Fill Site, 336
High Capacity Pumper, 345
High Capacity Pumper Tanker, 345
High Capacity Tanker, 344
High Capacity Water Shuttle, 322, 339, 344
High Capacity Water Tank, 345, 351
High Pressure Storage, 128
Hose Tests, 45, 400
Hose Thread Compatibility, 308
Hosehandlers, 289, 313
Hoseholder, 280
Hydrant Coverage Map, 35
Hydrant Gate Valve, 18, 380
Hydrant Layout Kit, 381
Hydrant Operation and Maintenance, 11, 46, 126, 137, 139, 142, 158
Hydrant Spacing, 35
Hydrant System, 10, 136, 294, 295, 296, 318
Hydraulics, 407
Hydro-pneumatic storage tank, 130

I

Initial Attack Lines, 11
Initial Fire Attack, 11, 59, 83, 380
Initial Water Supply, 2, 83, 85
Insurance Services Office (ISO), 32
Intake Relief Valve, 101
Interface Areas, 5
ISO Rating Schedule, 26, 34, 35, 41, 43, 45, 102, 104, 132, 158, 170, 206, 251, 351, 386, 407

J

Jet Dump, 263, 348, 368

L

Ladder Company, 41, 389
Large Diameter Hose, 20, 43, 78, 219, 302, 309, 411
Lateral Pipe, 188
Legal Liability, 49
Life Hazards, 6
Life Safety Code, 6
Lift, 90, 94, 173, 175
Loading Station, 306, 315
Long Range Budget, 50
Low Level Strainer, 44, 274, 277, 329
Low Pressure Storage, 131

M

Maneuvering Time, 264
Manifold, 271, 302, 310
Master Planning, 28, 48, 398
Master Streams, 89, 385
Mobile Water Supply Apparatus, 35, 63, 83, 104, 109, 117, 124, 391, 399
Modified Split Lay, 386
Multi-tank Dump Site, 324
Mutual Aid, 49, 66, 378, 399, 417

N

Needed Fire Flow, 3, 53, 165, 379, 398
Negative Pressure, 92
Net Pump Pressure, 92, 220
NFPA 1001, 49
NFPA 1231, 27, 29, 32, 52, 54, 104, 115, 379
NFPA 1410, 11, 65, 379, 386
NFPA 1500, 29, 49
NFPA 1901, 86, 173, 233, 360
NFPA 1903, 104, 344, 357, 360
NFPA 1911, 91
NFPA 291, 136, 155, 409

Nozzle Pressure Variations, 69
Nurse Tanker, 247, 267, 323, 387

O

Occupancy Hazard Classification, 55
Offensive Attack, 13
Open Relay, 235
Operating Budget, 50
Operating Pressure, 71, 220
Overflow, 361

P

Parallel Tank, 273
Personnel, 40, 49, 389
Petroleum Type Coupling, 263, 307
Portable Hydrant, 22
Planning, 30, 239, 324, 393, 396, 415
Pneumatic Tank, 130
Polypropylene Tank, 357
Portable Pump, 203, 205
Portable Tank, 44, 248, 269, 272, 367, 386
Positive Pressure, 92
Power Take Off (PTO) Pump, 365, 371
Power Transfer, 326, 329
Preconnected Lines, 83, 86, 87
Preplan, 66, 324, 338
Pressure Governor, 101
Pressurized Dump Lines, 271, 291
Pressurized Tank, 130
Primary Risk, 56
Primer, 100, 173
Private Water Systems, 397
Public Protection Classification (PPC), 26, 398
Pumper Outlet (Hydrant), 15
Pumper-Tanker, 105, 345, 374

Q

Quick Connect Coupling, 263, 308

R

RAWR, 108, 110
Relay Operation, 23, 212, 222, 228, 390
Relay Problems, 227
Relay Control Officer, 215, 413
Relief Valve, 101
Remote Loading Station, 309, 313
Remote Unloading Station, 291
Reserve Engine, 383
Reserve Pumper, 41
Reserve Tank, 326, 332
Reservoir, 130
Residual Pressure, 131, 220, 141, 157
Response Time, 379
Reverse Lay form Source, 21, 383
Rigid Frame Tank, 269
Risk Analysis, 10, 32, 54, 392, 394

S

Safety Officer, 291, 314
Safety Valve in Relay, 224
Service Company, 41, 389
Service Tests, 45, 94, 102, 104, 400
Siamese, 265, 387
Siphon, 326
Soil Conservation Service (SCS), 397
Special Risk, 10
Specifications, 376, 400
Spillage, 121
Split Hose Bed, 74, 385
Split Lay, 383
Staging Area, 315, 339
Standard Operating Procedures, 48, 228, 312, 378, 392
Standpipe, 128, 130, 204

Static Pressure, 128, 133, 141, 157
Static Pressure Loss, 188, 212
Static Source, 165
Storage Tank, 207
Storage at Dump Site, 266
Subsequent Alarm, 392
Suction Gated Wye, 222
Suction Siamese, 222
Suction Hose, 175
Suction Supply, 28, 93, 165, 167, 178
Sump, 359
Supplemental Suction Supply, 206
Supplemental Unloading Station, 336
Supply Lines, 72, 73, 75
Supply Tank, 326
Supply Works Capacity, 132, 167, 296, 386

T

Tank Capacity, 356
Tanker Coverage, 35, 405
Tanker Selection, 256
Tanker Routing, 258
Target Hazard, 48, 53, 398, 405
Testing, 85, 86, 90, 94, 96, 100, 104, 122, 124
Total Water Supply (TWS), 54, 57, 127
Traffic Control, 240, 259, 314, 316
Traffic Control Officer, 291
Training, 203, 235, 407
Training Records, 49, 261
Transfer Flow, 328
Transfer Pump, 329
Transport Capabilities, 45
Transport Distance, 388, 400, 401
Transportation Hazards, 7
Travel Time, 251, 339
Two Piece Engine Company, 212, 387

U

Unloading Station, 276, 291
Unloading Time, 263
Unprotected Risk, 28
Usable Water, 104, 114, 123, 267, 272, 288, 323

V

Vacuum Gauge, 177
Vacuum Reading, 176
Vacuum Tanker, 385
Vent, 359
Vent Crew, 290, 314

W

Waiting Time, 264
Water Delivery System, 81, 410
Water Department, 125
Water Fill Point, 316
Water Hammer, 138
Water Quality, 181, 183
Water Shuttle Operation, 24, 237, 244, 250, 260, 411
Water Shuttle Control Officer, 388, 413
Water Source Certification, 170
Water Supply Alarm, 392
Water Supply Availability, 62, 170
Water Supply Coverage Map, 46, 404
Water Supply Equipment, 43, 399
Water Supply Officer, 27, 215, 240, 242, 262, 315, 388, 393, 413
Water Supply Methods, 14, 410
Water Supply Point, 170, 211, 315, 394, 399, 409
Water Supply Preplan, 324
Water Supply Pumper, 83, 303, 387
Water Supply Tanker, 107, 346, 363
Water Supply System, 46, 125, 128, 133, 185, 262, 395
Water Tank Design and Construction, 350, 352
Weir, 167
Whirlpool, 178, 276
Wet Hydrant, 185